TERRESTRIAL ECOTOPIAS

Ralahine Utopian Studies

Series editors:
Raffaella Baccolini (University of Bologna at Forlì)
Antonis Balasopoulos (University of Cyprus)
Joachim Fischer (University of Limerick)
Michael G. Kelly (University of Limerick)
Tom Moylan (University of Limerick)
Phillip E. Wegner (University of Florida)

Volume 33

PETER LANG
Oxford - Berlin - Bruxelles - Chennai - Lausanne - New York

Heather Alberro

TERRESTRIAL ECOTOPIAS

MULTISPECIES FLOURISHING IN
AND BEYOND THE CAPITALOCENE

PETER LANG
Oxford · Berlin · Bruxelles · Chennai · Lausanne · New York

Bibliographic information published by the Deutsche Nationalbibliothek.
The German National Library lists this publication in the German
National Bibliography; detailed bibliographic data is available on the
Internet at http://dnb.d-nb.de.

A catalogue record for this book is available from the British Library.

Library of Congress Cataloging-in-Publication Data

Names: Alberro, Heather, author.
Title: Terrestrial ecotopias: multispecies flourishing in and beyond the
 capitalocene / Heather Alberro.
Description: Oxford; New York: Peter Lang, 2024. | Series: Ralahine
 Utopian studies, 1661-5875; volume 33 | Includes bibliographical
 references.
Identifiers: LCCN 2023056302 (print) | LCCN 2023056303 (ebook) | ISBN
 9781800795761 (paperback) | ISBN 9781800795778 (ebook) | ISBN
 9781800795785 (epub)
Subjects: LCSH: Green movement. | Utopias–Environmental aspects. | Human
 ecology. | Ecocriticism.
Classification: LCC GE195 .A434 2024 (print) | LCC GE195 (ebook) | DDC
 304.2-dc23/eng/20240131
LC record available at https://lccn.loc.gov/2023056302
LC ebook record available at https://lccn.loc.gov/2023056303

Cover image: Laura Parenti / Pexels.
Cover design by Peter Lang Group AG

ISSN 1661-5875
ISBN 978-1-80079-576-1 (print)
ISBN 978-1-80079-577-8 (ePDF)
ISBN 978-1-80079-578-5 (ePub)
DOI 10.3726/b18622

© 2024 Peter Lang Group AG, Lausanne
Published by Peter Lang Ltd, Oxford, United Kingdom
info@peterlang.com - www.peterlang.com

Heather Alberro has asserted her right under the Copyright, Designs and Patents Act,
1988, to be identified as Author of this Work.

All rights reserved.
All parts of this publication are protected by copyright.
Any utilisation outside the strict limits of the copyright law, without
the permission of the publisher, is forbidden and liable to prosecution.
This applies in particular to reproductions, translations, microfilming,
and storage and processing in electronic retrieval systems.

This publication has been peer reviewed.

Contents

Acknowledgements ... vii

Introduction ... 1

CHAPTER 1
The Deficient Present ... 11

CHAPTER 2
Supremacy ... 37

CHAPTER 3
A Better World is Reality ... 69

CHAPTER 4
Literary Ecotopias ... 109

CHAPTER 5
'The Great Refusal': Ecotopian Social Movements ... 143

CHAPTER 6
For the Love of Kin: Indigenous Green Futurisms ... 183

CHAPTER 7
For the Love of Kin: Terrestrial Ecotopias in and After the End ... 215

CHAPTER 8
Towards Multispecies Flourishing ... 245

Bibliography	253
Index	287

Acknowledgements

This work, though primarily based on the research that I undertook during my PhD, has been years in the making, crystallising especially after I first encountered the fields of science fiction and utopian studies during my undergraduate studies at the University of Miami. I'd like to thank my visionary and inspiring lecturers during those formative years for pointing me in the direction of radical authors and activists variedly challenging the status quo.

None of this work would have been possible without the ongoing support, encouragement and solidarity tirelessly displayed by my family and friends throughout the years: My mum Elaine Alonso-Cruz, my sisters Alexandra and Angela Cruz, my dad Jorge Cruz, my dearest friends and comrades James McIntosh, Alice Sibley, Pietro Gallanti, Rhiannon Firth, Nora Castle, Emrah Atasoy, Manuel Sousa-Oliveira, Athira Unni, Aiko Kurisutaru, Anja Wettergren, Tom Douglass, Luigi Daniele, Filip Nuyens, James Reid, Casper, and many more.

I also wish to thank colleagues and friends from the field of utopian studies – Tom Moylan, Laurence Davis, Darren Webb, Nicole Pohl, Ibtisam Ahmed, and many more. Since finding my disciplinary 'tribe', so much has fallen into place. The utopian 'lens' has helped me frame and articulate so many of our present injustices, and see that they are far from natural or inevitable. Thank you to Nottingham Trent University for supporting and variedly funding my research since 2016, and for the wonderful colleagues who've enriched my time at this institution.

A very warm 'thank you' to my friends and research participants from Earth first!, Hambacher Forst, Extinction Rebellion, SeaShepherd, hunt saboteur activist groups, and others for letting me gain further insights into the incredible work that you do. Your tireless efforts continue to inspire me.

To my wonderful, innumerable terrestrial kin, those with wings, scales and fur, those who crawl, fly, climb, run and swim; from the microscopic to the multi-tonned wonders … I love you, though I've only had the privilege of getting to know a few of you. I'm forever grateful to you for not only making 'me' possible, in a bio-physiological and psychological sense, but also for bringing such immense joy to my life since I can remember. This is for you.

Introduction

After living through the upheavals of two World Wars and the Great Depression, an ageing H. G. Wells, often referred to as the father of science fiction, appeared to lose much of his former, seemingly boundless optimism for the establishment of a peaceful and egalitarian future.[1] In his final work *Mind at the End of Its Tether* (1945), published one year before his death, he reflects on a chaotic and unpredictable war-torn world which has obliterated the 'orderly development of life' and thus the ability to 'sketch out the pattern of things to come' (5). In the 'voiceless limitless darkness' (15) that remains, Wells gleaned:

> the slowing down of terrestrial vitality ... [wherein] the human mind is active still but it pursues and contrives endings and death. The writer sees he world as a jaded world devoid of recuperative power. In the past he has liked to think that Man could pull out of his entanglements and start a new creative phase of human living. In the face of our universal inadequacy, that optimism has given place to a stoical cynicism. (30)

In a stark (and somewhat misanthropic) abandonment of his former hope in *human* perfectibility, Wells presages the imminent and inevitable end of the 'human story' (18) and life in general as we have known it, and gestures towards the emergence of a more adaptable posthuman species that shall inherit the earth (19). One can only speculate what Wells would opine if he were alive today, in a world plagued by crippling levels of socio-economic inequality, the disturbing proliferation of far-right social and political mobilisations (Lazaridis et al. 2016), a climate system in disarray, the accelerating absence of other-than-human life forms and the incalculable misery that is yet to come as the above unfold in tandem. Here it is of the essence to recall one of Wells's former admissions: that of the 'black, blank and vast ignorance' (Wells 1986, p. 120) of the to-come wherein

---

[1] In this final work Wells also alludes to a heart-related illness that likely further contributed to his emerging sense of finality and sombre reflections on the 'to-come'.

hope continually resides. With so much on the line, now more than ever we must recall that the 'Now' of late capitalism, with its myriad socio-ecological exclusions and systematic eradication of life's rich assemblages, is far from inevitable – it is merely one amongst innumerable possible worlds. That is what I will attempt to illustrate throughout this book. In this I follow Rebecca Solnit and others who, in her essential work *Hope in the Dark: Untold Histories, Wild Possibilities* (2016), urges that now more than ever – amidst the greatest unravelling of life's assemblages seen in millions of years, climate chaos, mounting socioeconomic inequality – we need hope. Hope functions as a life-saving light in the dark, one that refutes what is, whilst illuminating the other possibilities that lie around and ahead. However, crucially, hope is not reducible to mere wishful thinking or blind faith, nor 'a lottery ticket you can sit on the sofa and clutch, feeling lucky' ... hope is ...

> An axe you break down doors with in an emergency; because hope should shove you out the door, because it will take everything you have to steer the future away from endless war, from the annihilation of the earth's treasures and the grinding down of poor and marginal. Hope just means another world might be possible, not promised. (p. 4)

Neither the deficient present, nor a better world, is ever guaranteed; both are mere possibilities brought about and maintained through sensibilities, institutions, policies, mores, hegemonic values and innumerable micro and collective macro-interactions. Contrary to some of Wells's understandable yet misguided intimations at the end of his life, our planet's regenerative capacity is not exhausted. It always exceeds us, so that there is never an 'end' in an absolute sense. And, indeed, Wells suggests as much when he leaves the door open for the possible emergence of a new posthuman species. A better, more inclusive and regenerative world is most certainly possible, but we have to fight for it – collectively, with urgency and with everything we've got. Because, despite what the ideological maxims of global capitalism and its champions continually insist, this world is far from inevitable. At roughly 500 years old (Klein 2020), historically it is a relatively novel development within the wider trajectory of human affairs. Yet even in its relatively short lifespan, this system, its generation and exacerbation of, and confluence with, other oppressive structures – colonialism, white supremacy, patriarchy, the heteronormative regime – has plunged us into a cacophony of intersecting crises that bode decidedly inhospitable futures for the many on terra. Late capitalism continues to disavow our ancestors, erode our present and steal our future (Braidotti 2017, p. 160), enriching a tiny minority at the expense of the multitudes. Now more than ever, radical alternatives that resolutely proclaim the enduring possibility of better worlds – and tirelessly seek to enact them – are of the essence. We may not have a full consensus on where we wish to go, but what's increasingly apparent is that current conditions are inimical to the wellbeing of a considerable and growing portion of the earth's inhabitants.

Enter the perennial utopian impulse, which in its myriad manifestations, from literary texts to social movements, mobilises an active and militant hope in defiance of oppressive conditions and in service of the 'better'. In response to such intersecting crises, this book seeks to shed light on some of the myriad manifestations along the green or ecotopian spectrum. Specifically, this book explores and argues for the need to build an inclusive, terrestrial utopianism grounded on terra, in and beyond the 'end times' of climate and ecological breakdown. This modality, amongst other things, must be decolonial, actively engaging with others' perspectives on 'the good life' from a standpoint of humility and solidarity, and critically posthuman, relating to other species and terra as agentic persons who matter beyond their use value to humans. This modality further recognises the urgency and indispensability of crying a resounding 'No!' to anti-utopian and politically fatalistic narratives encouraging that we 'learn to die' in the Anthropocene (Scranton 2015). Rather, we must do precisely the opposite – we must learn to fight in the Anthropocene (Stolze 2018), resisting as well the resurgence of 'dystopia porn' (Singh 2018) that fetishises the purported unrealizability of better worlds. The recent proliferation of eco-dystopian and (post)apocalyptic films, series and fiction such as *Mad Max: Fury Road* (2015), *The Rain* (2018), *Snowpiercer* (2020) and their widespread appeal in popular culture likely reflects a burgeoning unrest over mounting ecological disintegration and uncertainty. Moreover, as discussed in Chapter 3, dystopias serve the vital critical function of showing us how bad things might get if we continue along our current trajectories and, through 'analytical and imaginative contestation' of these dark elements, seek to jolt us out of our complacency and thus spur transformative action (Moylan 2021, p. 99); However, we also need to critically engage with and direct our steps towards positive visions of better worlds, however fragmented, tentative and necessarily open-ended. That is the crucial function of utopia – seen in literary forms that vividly sketch the contours of possible alternative worlds, transgressive social movements vehemently resisting the deficiencies of the present and prefiguring better ways of relating to others, and in the 'lived utopias' of intentional communities[2] the world over as concrete exemplars of the possible (Sargisson & Sargent 2017). What's more, we must also resist the co-optation, domestication and commodification of the transformative utopian impulse effected by the logic of capitalist realism and its proliferating 'false novums' of commodified fantasies (Suvin 2000; Le Guin 2001; Webb 2016). Yet even here, hope springs eternal:

> Imagination like all living things lives *now*, and it lives with, from, on true change. Like all we do and have, it can be co-opted and degraded; but it survives commercial and didactic exploitation. The land outlasts empires ... (Le Guin 2001, p. xiv)

The open-ended terrain of potentiality abounds; the earth will outlast capitalism, though what kind of earth we inherit depends on what actions are taken today. What's more, long-oppressed groups (BIPOC, other species, etc.) have already lived through 'end times' imposed by colonialism, capitalism and human supremacy. Yet after 'the end' (Whitehead 2020), the boundless potentiality of life as such is what paves the way for new modalities to continually emerge. What ways of being do we wish to create together in the 'here below'? How do we ensure that we include our long-excluded terrestrial kin in our visions and strivings for more liveable futures? As the latter query in particular seems under-addressed in many ecotopian manifestations, this book is an attempt to map and sketch the contours of contemporary ecotopian visions for a multispecies flourishing. As I will argue throughout this book, the myriad crises we face require more

---

2   See the 'Global Ecovillage Network' for a list of thousands of such experiments in green utopian living around the globe.

than mere technical fixes but, crucially, fundamental ethical and political transformations, changes in how we perceive, value and relate to others, particularly other species. These crises require a radical rethink of how we define and conceive of 'the good life'. In so doing, we ought to continually ask ourselves: Who matters and why? How ought we (re)arrange society so as to enable all – humans and all terrestrials – to live the 'good life'? Who might be included as well as excluded in our (re)definitions of the 'good life'? Because Western culture has tended to value humans above all else, or more precisely white, heterosexual, wealthy, Christian, able-bodied men above all 'others'. The hierarchies and antagonistic dualisms inherent especially within western culture override the interconnections that always, inextricably embroil us in the lives of others, without whom we could not be. They blind us to the innumerable voices and agencies that continually surround us, compose us and sustain us. In surveying and critically analysing contemporary manifestations of ecotopianism in the chapters that follow, I ask in particular, whither the other-than-human? To what extent do these visions and strivings for ecologically resilient futures actively foreground the needs and wellbeing of our terrestrial kin rather than consign them to a mute backdrop overshadowed by a predominantly human drama?

In light of these critical and investigative endeavours, as well as my own research background, the following discussions proceed, not from an exclusive grounding in literary ecocriticism or other single disciplinary tradition, but from a thoroughly transdisciplinary perspective. That is, I variedly draw on allied disciplines at the intersections of the (critical) environmental humanities and social sciences, including but not limited to normative environmental ethics, literary ecocriticism, critical-posthuman theories, and decolonial political ecology. After all, the complex task of building better worlds is an inherently collective endeavour that requires contributions and insights from multiple lenses and positionalities. The structure of the book is as follows: Chapter 1, "The Deficient Present" will 'set the historical stage' by briefly cataloguing the myriad ecological dislocations of the 'Capitalocene' – namely climate breakdown and the sixth mass extinction – alongside key academic debates surrounding the origins and nature of this unprecedented epoch. The chapter will then look at critical debates around the 'Anthropocene' through additional 'boundary' concepts such as More's 'Capitalocene' and Haraway's 'Chthulucene'. The chapter concludes with reflections on the paradoxical nature of our present era, marked by the pervasive effects of human 'agency' and mounting retaliations by other terrestrials against (certain) human activities, and the need to radically rethink human – other-than-human relations in the challenging terrain ahead.

Chapter 2, 'Supremacy', serves as another foundational chapter, with a critical overview of the history of Western Anthropocentrism – or the paradigm of 'human exceptionalism – that I and many other scholars argue is a key underlying factor of contemporary ecological decline. I offer a brief discussion of how Anthropocentrism still underpins mainstream approaches, values and behaviours in relation to other species, including the discourses and policy approaches of major 'sustainable development' actors like the UNEP and WBCSD. Understanding the nature and driving factors of culturally dominant human-animal hierarchies and dualisms is crucial for gauging the extent to which ecotopian visions challenge and depart from such longstanding worldviews. I reiterate the ethical imperative of foregrounding our terrestrial kin in any conceptualisations of – and strivings towards – better futures.

In Chapter 3, 'A Better World is Reality', I introduce and expand upon two of the key theoretical tributaries informing this book, which I draw upon as lenses through which to critically examine contemporary ecotopian manifestations. I begin with a critical overview of some key developments, concepts, ideas and thinkers in relation to the discipline of utopian studies, and then critical-posthumanism and green utopianism in particular, followed by a discussion of the continuities uniting the two traditions. I close with reflections on the potential of both for helping us critique the present and radically reconfigure interspecies relations along more ethical and inclusive trajectories.

Chapter 4, 'Literary Ecotopias', looks at the first medium or manifestation of ecotopianism in the form of features a brief and critical overview of a selection of canonical along with more obscure ecotopian literary texts through the lens of critical-posthuman and green utopian theory. The central starting point is the books main investigative thread: 'whither the other-than-human?' From this foundation, I proceed to analyse how do the different texts portray other animals and beings in their visions of more socio-ecologically resilient futures. Specifically, I enquire, 'what role, if at all, do other species play in these texts – as mere backdrop to human-centered ecological sustainability concerns?' As active agents, or even central protagonists? To what extent are their unique needs taken into account in the texts' descriptions of ideal ecotopian societies? I conclude with reflections on the role of literature and literary ecocriticism more generally in the vital process of radical social, political and cultural change.

In Chapter 5, '"The Great Refusal": Ecotopian Social Movements', I draw on original empirical data gathered during and after my PhD investigations of the ecotopian potentiality of contemporary radical environmental mobilisations – i.e. Earth First!, Sea Shepherd, Extinction Rebellion, Ende Gelände, and some more recent mobilisations including Les Soulèvements de la Terre and JSO. The aim is to assess the ecological worldviews and ecotopian visions of these myriad groups, and – similarly to the previous chapter – the extent to which they transgress Anthropocentric hierarchies and dualisms in their reconfigurations of interspecies relations in their worldviews and visions for more sustainable futures. I also explore the unique modality of ecotopianism exhibited by these groups and the conditions shaping it. At the outset I encountered considerable and unforeseen methodological challenges. I was advised against my initial plan of conducting field research aboard one of the Sea Shepherd vessels by my supervisory team and university ethics committee. This months-long venture would likely have involved considerable risks from encounters with pirate fishing and whaling vessels. Moreover, building trust and rapport with participants from groups such as EF! ended up being a lengthy and challenging process due to their understandable scepticism towards newcomers born of repeated encounters with undercover police over the years. Partly due to such issues surrounding access, I largely abandoned face-to-face interviews in favour of synchronic online interviews (McCoyd & Kerson 2006, p. 390; Deakin & Wakefield 2014). No notable differences in the quality of interviews obtained were detected; in fact, a range of benefits including greater assurance of participant anonymity were noted. The final sample featured semi-structured interviews with 26 REAs affiliated with groups such as Earth First!, Sea Shepherd, Hambacher Forst, Hunt Saboteur groups and Extinction Rebellion. The interviews, from which extracts are featured below, were conducted primarily via online platforms such as Skype with activists across the UK and Western Europe, Canada, Australia, North and South America. All data was fully anonymised through the use of pseudonyms in order to conceal participant identities. Organisational documents (Bowen 2009) were also utilised as supplemental data in support of data obtained from the semi-structured interviews. Lastly, I utilised thematic analysis (Braun & Clarke 2013) for developing, contrasting and analysing common threads across the varying data sets.

Chapter 6, 'Good Relatives: Indigenous Green Futurisms', is an attempt to continue the vital work of decolonising the utopian canon by exploring indigenous cosmologies, ethics and how they relate to indigenous ecotopian mobilisations and conceptualisations of 'the good life'. I examine some examples of indigenous onto-epistemological and ethical modalities in relation to the other-than-human world, how the former influence their approaches to u(eco)topianism, and conclude with a discussion of Joshua Whitehead's wonderful edited collection of indigiqueer speculative fiction *Love After the End* (2020).

In Chapter 7, 'Terrestrial Ecotopias in and After the End', I aim to unite and reflect on some of the key themes developed throughout the book. I reflect on the nature and significance of the widespread loss of our terrestrial kin during the present, and why (green) utopian imaginaries are especially important as a counterforce and way of generating radically different ways of living well with other terrestrials in and beyond the Capitalocene. I then discuss in greater detail the politics of conviviality, and include a brief overview and discussion of strands of ecotopianism in mainstream developments, policies and proposals. I offer critical reflections on notable recent developments such as various global iterations of the 'Green New Deal and urban and rural rewilding projects around the world. The aim will be to shed critical light on these various concrete examples of attempts to 'build back greener', their limitations and ecotopian potential.

In the brief and final chapter (Chapter 8), 'Towards Multispecies Flourishing', I flesh out what I refer to throughout the book as a 'terrestrial utopianism' that is alive in radical texts and movements in the present. It is this modality of ecotopianism, which is rooted in the earth and in-relation with other species, that I maintain holds particularly radical potential for building more liveable worlds in and beyond the Capitalocene.

CHAPTER 1

The Deficient Present

Paroxysms and Perturbations of a Planet in Peril

As we will see in Chapter 3, utopian strivings in all of their manifestations have two key components: staunch criticisms of a deficient present and imaginative projections of better worlds. Thus, a book on contemporary ecotopianism and the role of our terrestrial kin in them must first begin with a brief discussion of the socio-ecological deficiencies of the present, what so many scholars, activists and Indigenous protectors the world over are fighting against. It is no secret that we find ourselves amidst a litany of unprecedented and worsening social and ecological crises, to the extent that loss and disintegration appear to be defining features of our contemporary era. Hence, the now well-worn term for the sheer extent of the impacts of modern societies on the earth's natural systems and other terrestrial life forms: the Anthropocene (Crutzen 2006), the age of unprecedented, ubiquitous and destabilising anthropogenic pressures on the earth system (Hamilton 2016). The Anthropocene concept denotes a fundamental change in the relationship between humans and 'nature', one so significant as to mark the dawn of a new epoch wherein one single entity has become a global 'geophysical force' (Steffen et al. 2007). Below, I will return to this concept and discuss some of its limitations. For now, suffice it to say that despite issues with what the Anthropocene narrative designates as the key drivers of contemporary ecological breakdown, the upheavals that it casts a spotlight on warrant close and immediate attention.

As I write, the world has warmed over 1 degree Celsius since the industrial revolution, one of the key watershed moments that altered the course of modern history in myriad and fundamental ways. The effects of this seemingly insignificant degree of excess heat in a system as complex as the climate are already painfully apparent. The recent catastrophic floods in Pakistan which displaced over 33 million people and destroyed millions of acres of agricultural land (UNICEF 2023) are but the latest reminder that climate change is no threat looming on a distant horizon, but is already here, and indeed has been so for some time, especially for the world's most marginalised populations. In this rapidly warming world, with climatic changes that continue to outpace expert predictions, the decades ahead will see more violent storms, raging wildfires, prolonged droughts, rising food insecurity due to crop failures, millions of people – human and other-than-human – on the move in search of habitable dwelling places, and intensified geopolitical conflicts. The first segment of the IPCC's Sixth Assessment Report (2022) warned that without 'rapid and deep' emissions reductions through fundamental transformations of everything from the housing sector to energy, transport and agriculture, 'GHG emissions are projected to rise beyond 2025, leading to a median global warming of 3.2 [2.2 to 3.5] °C by 2100' (p. 21). This is a frighteningly far cry from the upper limit of the Paris Agreement's objective of limiting warming to under 2 degrees by 2100. A 3-degree world would be nothing short of apocalyptic for much of life on earth. To put the situation into perspective, one recent international study of climate-related mortality suggests that up to an additional five million lives were lost annually between 2000 and 2019, a period which saw a 0.26C-rise per decade, due to extreme temperatures (Monash University 2021). The World Health Organisation (WHO 2021) predicts an additional 250,000 additional deaths *per year* [emphasis added] between 2030 and 2050, at the conservative end, from health-related impacts of runaway climate change.

The Era of Global Boiling

The final segment of the IPCC's Sixth Assessment Report (2023), a compilation of eight years' worth of investigations of anthropogenic climate change, has issued a 'final warning, urging swift and radical action in order to stem irrevocable socio-ecological chaos. In addition to implementing radical transformations across multiple sectors, predominantly in industrial-capitalist societies, the report emphasises that any new fossil fuel projects would be utterly incompatible with keeping warming below 2 degrees by 2100. It's important to highlight that the warming effects of rising greenhouse gas emissions are cumulative. Long-term and end-point targets such as reaching 'net zero by 2050' 'have no scientific basis'. 'What governs future global temperatures and other adverse climate impacts are the emissions from yesterday, today and those released in the next few years' (Anderson & Bows 2012, p. 639). In other words, pledging to reach carbon neutrality by 2030, 2050 and beyond amounts to far too little, too late. And yet, such long-term targets remain key pillars of global climate politics. And even by its own standards, the global community is falling abysmally short. According to the Climate Action Tracker, which ranks countries' climate mitigation plans (39 countries in addition to the EU, covering 85% of global emissions) from '1.5C Paris Agreement compatible' to 'critically insufficient', as of August 2023 most countries' pledges fell within the 'Highly insufficient' and 'Insufficient' categories (Climate Action Tracker 2023a). None were compatible with the Paris Agreement's more ambitious target of limiting warming to 1.5C. In an almost comical farce, the UAE, which is set to host COP28, still plans to increase fossil fuel production and consumption (Climate Action Tracker 2023b). Such accumulated neglect has already begun bearing fruits: in early July 2023, countries around the world recorded their hottest days ever – for four days in a row – with Death Valley, California, purportedly reaching an unimaginable 56.7C. The new global average temperature of 17.18C, also shattered previous records, with resulting wildfires and other heat-related casualties. July 2023 shattered precedents as the hottest month ever recorded in human history, prompting the UN Secretary General Antonio Guterres to herald the onset of the era of 'global boiling' (Niranjan 2023).

A provocative 2012 article by climate scientists Professor Kevin Anderson and Professor Alice Bows made a sobering assessment of 2030 and 2050 emissions targets put forth by nation states around the world in order to meet Paris Agreement targets. Because of the cumulative nature of emissions, long-term reduction targets actually have no grounding in climate science. What affects future warming trends is what we did *yesterday*, and what we do *today*. Pledging to reduce emissions to net zero by 2025, 2030 and beyond amounts to too little, too late. Even more striking is the article's bold – and decidedly utopian – conclusion that even achieving a 2-degree warming cap is likely impossible 'within orthodox political and economic constraints' and that scientists need to 'have the audacity to think differently and conceive of alternative futures' (Anderson & Bows 2012, p. 640). Indeed, ours is a crisis of imagination as much as it is of mounting greenhouse gases and collapsing wildlife populations. And there's another problem with long-range emissions targets: they evoke a sense of the climate crisis as something far away, still looming on the horizon. Yet the effects of climate breakdown, especially for the world's most vulnerable populations, arrived some time ago. The IPCC's third working group of its sixth assessment report (2022) contains even starker warnings – without 'rapid and deep' emissions reductions through fundamental transformations of everything from the housing sector to transport and agriculture, 'GHG emissions are projected to rise beyond 2025, leading to a median global warming of 3.2 [2.2 to 3.5] °C by 2100' (p. 21). Crucially, it warns that weather and climate extremes have already led to some irreversible impacts, with natural and human systems increasingly being pushed beyond their ability to adapt. Yet the report, too, leaves the door of possibility open, emphasising that there are a range of possible climate futures that could yet emerge as well as be averted depending on how quickly, and to what extent, we act.

Epochal Transformations

Decades before choking in anguish in 1973 upon witnessing the extent to which his beloved dunes had been swallowed up by Rotterdam Harbour's steady expansion, wherein he'd painstakingly studied gulls, terns, and various bird species, Dutch biologist Nikolaas Tinbergen issued the haunting omen: 'it will all go, irrevocably' (de Waal 2016). Recent research has found that (industrial) humanity has severely altered 97% of the earth's land, resulting in varying degrees of diminished ecological integrity (Plumptre et al. 2021). Forests, especially rich terrestrial biodiversity, housing 60,000 different tree species, 80% of amphibian species, 75% of bird species, and 68% of the world's mammal species (FAO 2022). Around 10,000 years ago, forests covered around 57% of the earth's land surface; today they cover around 31% (FAO 2022). Commercial agriculture – i.e. the production of commodity crops such as beef, palm oil and soya (largely to feed cattle) – accounts for over 90% of tropical deforestation (Chalmers University of Technology 2022), much of which is done in order to satisfy demand for meat in Western Europe and China. A sobering recent study found that more than 75% of Amazon rainforest has been losing resilience since early 2000s due to climate change and other anthropogenic pressures (Boulton et al. 2022). The authors caution that the Amazon forest is nearing a 'critical threshold of forest dieback' with 'profound implications for biodiversity, carbon storage and climate change at a global scale' (Boulton et al. 2022, p. 271). The razing of forests and other flora has become so severe that some have documented a twofold decline in plant biomass since the start of human civilisation (Bar-On et al. 2018), with over half of the world's tropical forests destroyed since 1960s alone (IUCN 2021). Though rates of deforestation are declining, around 10 million hectares of forest were lost annually between 2015–2020, accounting for around 420 million hectares lost during this period, an area the size of Algeria and Libya combined (FAO 2022). Such losses further lead to disastrous impacts on the climate through CO_2 release, disruptions to terrestrial biogeochemical and water flows, and habitat fragmentation resulting in further biodiversity loss. Industrial humanity's collective Ecological Footprint, a function of both population size and particularly consumptive profligacy, has been exceeding the planet's carrying capacity since the 1970s. At present the resources demanded by the 'explosion of human consumption' as seen in the near exponential increase in the demand for energy, water, and land require the regenerative capacity of nearly two planet earths for their continuation (Global Footprint Network 2022). Though, as I'll discuss below, the Global Footprint Network's framing of a homogenous 'humanity' as exhausting the earth's resource budget prematurely overlooks substantial inequalities in consumption trends and resulting emissions linked with socioeconomic status. An entire planet with the same energy and resource-intensive lifestyles as average denizens of wealthy industrial nations would be inconceivable; Although, one can hardly blame long-struggling developing nations for desiring to do so, given that the Global North has historically and presently enriched itself at their expense (Agarwal & Narain 2017).

Fossil fuel extraction is the second most significant driver of environmental degradation via GHG emissions (especially CO_2 and Methane CH_4, which is thirty-four times more effective at trapping heat than carbon dioxide) from fuel combustion for electricity, heat, and transport. In the fossil industry's desperate scramble for the last remnants of increasingly scarce oil reserves, we've seen the emergence of particularly disastrous (socially as well as ecologically) unconventional fuel sources and extraction techniques such as hydraulic fracturing and horizontal drilling (fracking) (Willow & Wylie 2014). If we are to have any chance at limiting planetary warming to 1.5 degrees, the lower benchmark of the Paris Agreement, let alone 2 degrees, CO_2 emissions must remain below a maximum of 1,240 giga-tonnes (Gt) by 2050; this would require leaving the majority of known fossil fuel reserves – which contain an estimated 11,000 Gt of CO_2 – untouched (Jakob & Hilaire 2015) – in the ground. Otherwise, climate-related disruptions such as droughts, floods, and famines portend mass social dislocations the world over, with rising sea levels alone set to displace tens of millions worldwide. In 1992, the nongovernmental organisation, the Union of Concerned Scientists, gathered 15,575 of the world's most eminent scientists to issue a report entitled 'World Scientists' Warning to Humanity'. The report featured pleas for a radical overhaul in human production and consumption patterns which were then already posing severe threats to the planet's life support systems. It noted that if fundamental changes weren't made such as more equal resource distributions, stabilisation of human population growth, and a vast downscaling of environmentally destructive activities, serious risks would be posed to the future of life on earth (UCS 1992, p. 8). Three decades later, they've issued a 'Second Notice' denoting that the situation has severely worsened, especially with regards to climate change and biodiversity loss. Though they uncritically place primary emphasis on human population growth as a key driver of ecological decline, they further cite our collective failure to 'reassess the role of an economy rooted in growth' (Ripple et al. 2017, p. 1026) as another key factor. Thus, the growing consensus is that our presently 'failed' socioeconomic trajectories must be radically reconstituted, and that time is quickly running out for avoiding mass misery for much of life on an increasingly inhospitable planet.

Yet a cavernous gap seemingly persists between the urgency of the climate crisis and the lack of widescale concern and mobilisation on behalf of general populations and, in particularly, many in positions of key political influence. Though due to a range of complex political, cultural, socioeconomic and other factors, philosopher Timothy Morton (2013) posits that one possible explanation is that climate change can be seen as a 'hyper-object'. Hyperobjects have such a vast extension across time and space that they're historically beyond the range of human cognition; in other words, they 'massively outscale us' (Morton 2013, p. 12). This is undoubtedly further compounded by established interests desperately trying to constrict the horizons of political possibility in order to ensure their continued profitability. For instance, in the US polarisation on climate views can be linked to, among other things, systematic efforts to spread doubt about the reality of climate change through conservative media platforms (Carmichael & Brulle 2017). The present conservative UK government, former host of COP26, has invested £20bn more in support of fossil fuel projects than in renewables since 2015 (Horton 2023). Similarly, Prime Minister Rishi Sunak recently approved 100 new licences for oil and gas exploration in the North Sea, a move that he, in a seeming bout of madness, deems consistent with the Paris Agreement's Net Zero targets. Relatedly, what do we call the recent findings by the Center for Climate Crime and Climate Justice based at Queen Mary University of London (Whyte 2023): that shareholders in BP and Shell have purportedly earned 'a total of £131 billion in dividends and share buybacks combined since the Paris Agreement was signed', and that in this same period, the top eight shareholders have 'significantly expanded their holdings' in BP and Shell? The figure above merely represents the value of cash earnings and not that of their shares, which also rose significantly during this period. As we'll see in Chapter 5, 'ecotopian' social movements including the 'youthquake' of young peoples' environmental mobilisations across the globe as evinced by Fridays for Future and the Sunrise Movement (Sloam et al. 2022), are keeping the radical hope for thoroughly transformative change alive. Moreover, recent polls conducted in the UK have shown that the general public is overwhelmingly in favour of tougher action on climate change, and thus similarly supportive of non-violent direct-action tactics in service of such action (Carrington 2022).

Kin on the Brink: The Sixth Mass Extinction

> What a gap in the world, the missing white frost-face of that slim mountain lion.
>
> – Lawrence in Muldoon (1997, p. 158)

The notable French science fiction writer Jules Verne's classic work *Twenty Thousand Leagues Under the Sea*, first published in 1870, contains breathtakingly vivid and detailed descriptions of the incredible diversity of our marine counterparts. The novel is also rife with foreboding reflections on the likely irreversible decline of many species should over-hunting and similar anthropogenic pressures continue unabated (2017, pp. 361, 387). Verne worried that the 'barbarous, unthinking ferocity of whalers will one day obliterate the last whales from the ocean' (2017, p. 361). Thankfully, in this instance, after mounting social movement and public pressures, and global cooperation via the establishment of the IWC in 1946, some species like the humpback whale are making a comeback (Zerbini et al. 2019). Three decades ago, French sociologist André Gorz (1987) enquired, 'Did you know that, according to Cousteau, half of the marine life he filmed in 1956 had disappeared by 1963 (and what is left today)?' (p. 64). The overall situation today, decades after Verne's (2017) and Gorz's (1987) fateful observations, is in many respects scarcely imaginable. Over the last half-century in particular, mounting socioeconomic activity spurred by high-consuming industrial nations has been associated with the gradual simplification, range-reduction, and disappearance of biodiversity and especially megafauna populations across the globe (Ripple et al. 2019). Many others have keenly, and painfully, observed the gradual disappearance of other-than-human life forms throughout the post-war era, especially since the 1970s, and attempted to articulate it in different ways. Biologist Norman Myers referred to it as the 'great dying' in his pioneering and aptly named work on the subject, *The Sinking Ark* (1979). British naturalist Michael McCarthy in his heart-wrenching book *The Moth Snowstorm* (2015) recounts a childhood moulded by an abundance of lifeforms, wherein moths were so plentiful that they would overrun car headlights 'like snowflakes in a blizzard' (p. 13). He refers to the steadily creeping absence of life that has gathered momentum over the last few decades, wherein in many places to see a moth is an increasingly remarkable occurrence, as 'the great thinning'. Similarly, in his poignant work *Why Look at Animals?* (2009), English novelist and art critic John Berger produced a poem, 'They are the last', in an attempt to come to grips this troubling new phenomenon:

> Each year more animals depart. Only pets and carcasses remain, and the carcasses living or dead are from birth ineluctably turned into meat ... Once the animals flowed like milk. Now that they have gone, it is their endurance that we miss ... (p. 77)

An article by American herpetologist David Wake and researcher Vance Vredenburg documented precipitous global declines in amphibian populations (i.e. frogs and salamanders) as early warning signs that something was seriously amiss (Wake & Vredenburg 2008). The tragic irony is that amphibians, like sharks, are among the most ancient vertebrates the Earth has ever known, having emerged hundreds of millions of years ago and survived five previous extinction events; now amid the Anthropocene, they have become the world's most endangered class of animals. In their more recent analysis of current rates of species loss in comparison to those seen during previous mass extinction events on Earth,[1] Ceballos et al. (2020)

---

1 Such as the devastating Permian-Triassic event around 250 million years ago which wiped out 95% of all living species (Wake & Vredenburg 2008).

have observed an 'extremely high degree of population decay' in countless vertebrate species that is at least hundreds to thousands of times higher than 'natural' background rates of extinction, offering irrefutable evidence of a sixth mass extinction. In May 2019, a landmark and unprecedentedly comprehensive report from the Intergovernmental Science-Policy Platform on Biodiversity and Ecosystem Services (IPBES) uncovered the truly dire state of life on earth, with approximately 1,000,000 known animal and plant species threatened with extinction. Three years later, the WWF's latest 2022 Living Planet Report was released, utilising the findings from such indices as the Living Planet Index (LPI), which measures the global state of biodiversity and ecological health via an extensive compilation of over 3,000 data sources monitoring nearly 32,000 populations of vertebrate species (mammals, birds, fishes, amphibians, reptiles) across the globe (WWF 2022). The report denotes that due especially to land-use changes largely for agricultural expansion, we have seen a staggering 69% decrease in the abundance of monitored wildlife populations between 1970 and 2018, although such declines stem further back and have accelerated especially during the post-war era.

Many shark and ray species have experienced especially precipitous declines – approximately 71% between 1970 and 2018 – due primarily to an eighteen-fold increase in fishing pressures over this time period (WWF 2022, p. 42). Whilst comprehensive measures of the state of the world's invertebrates tend to be comparatively limited, a recent study documented 'catastrophic' declines in global entomofauna (insects), with nearly 40% threatened with extinction; particularly vulnerable are species in McCarthy's beloved Lepidoptera taxa, which includes butterflies and moths (Sánchez-Bayo & Wyckhuys 2021). Many have pointed out that what makes the present era of biological annihilation or 'defaunation' (Dirzo et al. 2014) unique is that 'a preponderance of evidence' suggests that rather than resulting from some external threat or cataclysmic climatic event, this time a 'single species' – humanity – is the leading cause (Cafaro 2015; Young et al. 2016; Bar-On et al. 2018). This point requires further, critical interrogation, and will be returned to in the sections below. As conservation scientist Gerardo Ceballos (2016) laments, 'we are in the midst of a massive assault on living things, causing the loss of millions of populations and thousands of species', species which are 'our companions in our travel across the universe' (pp. 290, 291). Ceballos (2016) asserts that this is no mere existential crisis threatening the continuity of human life but, fundamentally, an ethical one implicating the steady erosion of intricate multispecies relations forged over vast swaths of evolutionary time. The extension of the human carrying capacity has come not merely at the hands of technological innovation and more accurate manipulation of natural processes but, most insidiously, via the expropriation of ever larger portions of the Earth through the systematic displacement and annihilation of other beings (Crist 2017). Indeed, today in place of a diverse mosaic of life forms, humans and livestock have increasingly taken centre stage, with a collective biomass that now exceeds that of all other vertebrates apart from fish (Bar-On et al. 2018). The collective biomass of the planet's endlessly diverse insects, on the other hand, most of whom we know virtually nothing about, is in near freefall (Goulson 2021).

It has, in multiple senses of the phrase, become a largely human planet marked by a steadily growing absence of other life forms and ever fewer spaces not thoroughly (and dangerously) altered by 'us', or more specifically as I'll discuss subsequently, the necropolitics of late-stage capitalism. The WWF Living Planet Report 2022 urges that 'transformational' changes across every aspect of modern societies, as well as in terms of rights and how we value 'nature', are essential for mitigating the biodiversity 'emergency' (p. 4). And this is no easy feat, for issues of socio-ecological conflict and decline are exceedingly complex. For instance, regarding species over-exploitation, a major driver of biodiversity loss – the plight of the pangolin, the world's most trafficked mammal, is a case in point. Tracing the commodity chains of the international pangolin trade reveal a confluence of contributing factors – i.e. cultural beliefs in African countries like Ghana and Nigeria which tend to view wildlife as bushmeat, and economic factors such as rural poverty and foreign market demand which make the highly prized pangolins a difficult offer to pass up for hunters and traders simply trying to feed their families (Bachmann et al. 2019). This is why being quick to blame humans as a whole for the decline of other species is far too simplistic and dangerously misleading. Moreover, complex though these issues are, and though much has been lost, the present, widespread unravelling of terrestrial assemblages need not be so – another world more conducive to the flourishing of life is possible. Indeed, the WWF 'Living Planet' website has a tab entitled 'A different future is possible' delineating a range of potential solutions for turning the tides of loss. As I'll explore in subsequent chapters, such myriad deficiencies are what many contemporary forms of green utopianism have arisen in response to and in vehement defiance of. Before this, however, what of the origins of this purportedly new 'Anthropocene' epoch? Who is the implicit 'we' to which the term 'Anthropocene' refers and assigns culpability for our contemporary socio-ecological woes?

The Anthropocene: A Homogenous *Homo Sapiens* as Primary Earth Movers?

> The immemorial, fixed Earth, which provided the conditions and foundations of our lives, is moving, the fundamental Earth is trembling.
>
> – Serres (1995, p. 86)

Though there is a growing consensus that we are in a novel epoch, competing theorisations surrounding the precise origins, nature, and implications of this unprecedented transformation abound. Experts variedly point to the mutually reinforcing effects of unfettered (and grossly uneven) economic growth (Gorz 1987), growing populations, and rapacious production and consumption patterns particularly on behalf of the planet's plutocratic elite (Piketty & Chancel 2015). The French philosopher Michael Serres (1995) notably suggested that a defining feature of the Anthropocene is not merely the ubiquitous and destabilising influence of human activities by which the Earth *is moved* but, crucially, the mounting realisation that the Earth *moves* in response. In other words, as we'll explore later and in subsequent chapters, the new epoch is rather paradoxical, marked not only by the consequences of a destabilising and seemingly ubiquitous human agency but also by forceful retaliations by other entities to those actions. Some (Broswimmer 2002) have sought to associate the beginnings of the 'Anthropocene' with the development of human language and conscious intentionality. However, this problematically suggests that our contemporary destabilisation of Earth systems is inevitable rather than the result of particular historical, socioeconomic and political developments. Steffen et al. (2011) and Glikson (2013), on the other hand, contend that even prior to the development of language came modern humans' discovery and manipulation of fire. This, so the idea goes, was purportedly the 'evolutionary trigger' that allowed us to tap the powerful energy reserves stored in detrital carbon (Raupach & Canadell 2010, pp. 210–211),[2] effectively transforming *homo sapiens* into bio-cultural beings with increasingly sophisticated weaponry, modes of organisation, and a growing capacity to greatly alter our surrounding environments. These developments, so suggests this Anthropocene origin story, in turn engendered a dramatic shift in our relationships to the natural world and our terrestrial counterparts (Dawson 2016, pp. 20, 21). Still others have sought to locate the origins of the Anthropocene around the Pleistocene megafauna extinction event that took place around the last ice age. Herein, mass die-offs of numerous large mammals across Asia, Australia, and the Americas coincided with the arrival of early humans on these continents (Steffen et al. 2011, p. 847).

All living organisms necessarily impact their surrounding environments to varying degrees. Amongst humans, many ancient and pre-industrial civilisations often effected large-scale environmental transformations and were thus by no means ecologically benign. For instance, pre-industrial civilisations tampered with the energy-rich fossil fuels that now power modern industrial societies, most notably China's Song dynasty which made use of coal between 960 and 1279 (Wright 2012). However, the impacts of such developments were relatively minor in comparison with the monumental transformations that followed the Industrial Revolution and associated large-scale extraction of fossil fuels from the Earth (Crutzen 2006). The fundamental test of the Anthropocene is whether human activity affects earth

---

2   Recall Prometheus, a Greek mythological character and a former Titan who stole the 'red flower of fire' and gave it to humanity. With this newfound 'technology' humanity 'dug iron out of the earth...made weapons that were sharper than the lion's teeth; he tamed the wild cattle by fear of it' (Hyde 1905, p. 3).

system functioning to the extent that such impacts lie *outside the range of natural variability* (Steffen et al. 2015; Hamilton 2016) through, for instance, producing notable stratigraphic signatures in ice and sediments (Waters et al. 2016). The now widely established periodisation of the Anthropocene around 1800 following the momentous productive transformations ushered in by the Industrial Revolution (Steffen et al. 2011, p. 849) was initially proposed by Crutzen (2006; Crutzen & Stoermer 2000) precisely because this is when ice core data reveal considerable spikes in global concentrations of CO_2 and Methane (CH_4) arising from the industrial-scale burning of fossil fuels (Hamilton 2016, p. 2). In 1850 atmospheric CO_2 concentrations were at 285 ppm (parts per million), at the upper reach of natural Holocene variability of 260–285 ppm. By 1900, this number had climbed to 296 ppm, denoting an 'unmistakable human imprint' (Steffen et al. 2011, pp. 848–849) on the earth's atmosphere. From this moment onwards, humans developed the capacity to alter the very chemical composition of the earth's atmosphere (Steffen et al. 2011, p. 846).

Of course, humans are not the first or primary Earth movers; organisms such as cyanobacteria, phytoplankton and viruses have long since possessed Earth-moving capabilities, as has been made painfully apparent by the COVID-19 pandemic. As Donna Haraway (2015) observes, bacteria and their kin have always been and remain the primary terra-formers, though anthropogenic agency has become a dominant contemporary force. Since the 1800-marker, a new periodisation has taken centre stage: the post-1945 era of mounting socio-ecological upheaval known as the 'Great Acceleration' (Steffen et al. 2015). The Polish geologist Jan Zalasiewicz and colleagues (2015) notably suggest that the world's first nuclear bomb explosion, on 16 July 1945 at Alamogordo, New Mexico, marks the Holocene-Anthropocene boundary. This is rather fitting considering that from this period onwards there appears an 'explosion' of the 'human enterprise', particularly through near exponential increases – especially in the Global North – in consumptive and productive activities especially after the 1970s (Steffen et al. 2007, 2015). The post-war era of mass production and consumption in the form of soaring world trade and GDP has led to stratigraphic deposits that now contain novel materials such as plastics and concrete (Waters et al. 2016). Human bioturbation ('anthroturbation'), or significant disturbances of landscape, soil morphology and earth sediments resulting from such activities as deep mining for mineral extraction and fracking, has far exceeded that of any other organisms and placed unprecedented pressures on keystone bioturbators such as earthworms (Zalasiewicz et al. 2014). It appears that no place, not our food (Karami et al. 2017), our bodies (Anbumani & Kakkar 2018), nor even the earth's crust is devoid of the ubiquitous presence of 'industrial humanity'. However, as discussed below, these repeated emphases on a universal 'humanity' conceals myriad structural and intersecting inequalities that in turn fuel socio-ecological decline. Moreover, the politically fatalistic implication and at times outright explicit assumption that a universal humanity is at fault, or that somehow humans are an inevitably destructive species, leaves little room for effecting the fundamental societal transformations that are needed as a matter of urgency.

The Capitalocene

> We recognize that differential relations to power mean that not all peoples or species can equally access the possibilities contained in the future itself.
>
> – Kirksey and Chao (2022, p. 14)

Helpful as it has been in calling attention to and tracing key historical developments underlying the emergence of unprecedented global socio-ecological perturbations, the Anthropocene narrative has serious pitfalls in some of its foundational and diagnostic assumptions. An important point of departure is the common claim in dominant Anthropocene discourses encapsulated by environmental philosopher Holmes Rolston's (2015) assertion that for the first time in the history of life on terra, one single species can 'jeopardize the planet's future' (p. 32). Similarly, in their research on amphibian declines, David Wake and Vance Vredenburg note that, while multifactorial, the primary driver of amphibian collapse is 'one weedy species, *Homo sapiens*, which has unwittingly achieved the ability to directly affect its own fate and that of most of the other species on this planet' (2008, p. 11472). Many (Malm & Hornborg 2014; Alvater 2016; Moore 2016, 2017) have rightly problematised the totalising 'we' implicitly and explicitly called upon in such species-level framing. For instance, Ferdinand (2021) opts instead for the term 'plantationocene' (Haraway 2015) marked by the plantation at the 'center of the colonial inhabitation of the world'; this violent, racist and patriarchal system radically overturned pre-1492 ecosystems (p. 38) by transforming the land from that which nourishes peoples who inhabit it to something to be enclosed and heavily exploited largely for the enrichment of a few local colonists and overseas shareholders (pp. 38–40). So, the problem is not the rise of agriculture or the 'Neolithic Revolution' *per se*, but specifically of plantation, industrial,[3] and commercial-style agriculture. This differs starkly from, for example, *conuco* agriculture practised by pre-Columbian Amerindian groups like the Arawaks, wherein crops such as sweet potatoes and manioc root were cultivated together with care so as to prevent soil erosion (Ferdinand 2021, p. 39).

This critical counter-current gestures towards the structural – and colonial-capitalist – roots of contemporary socio-ecological disintegration. Jason W. Moore (2017) similarly locates the origins of our socio-ecological woes in capitalism's early modern period, beginning with Columbus' conquest of the Americas and the 'epochal transitions in landscape transformation after 1450', which marked the greatest watershed since the rise of agriculture and the first cities (p. 596). The year 1610, to be precise, saw the widescale reforestation of the American continent – and subsequent mass sequestration of CO_2 from the atmosphere – following the systematic extermination of the region's native populations (Lewis & Maslin 2015). Moore points out that the radical reshaping of global natures ushered in by colonial-capitalism and its violent dispossessions and land grabs caused social and ecological havoc long before the invention of the steam engine (p. 596). Referring to the current epoch more accurately as the 'Capitalocene', Moore (2017) challenges the Eurocentrism embedded within predominant Anthropocene discourses, posing the poignant query: 'Are we really living in the Anthropocene – the "age of man" – with its Eurocentric and techno-determinist vistas? Or are we living in the Capitalocene – the "age of

---

3   This aspect, of course, is not exclusive to colonial-capitalism but also was a central pillar of the disastrous Soviet-style '5-year plan' (Solnit 2021).

capital" – the historical era shaped by the endless accumulation of capital?' (p. 596; Žižek 2011). Colonialism and capitalism expanded their power in tandem from the sixteenth century as European powers plundered distant people and places under the pretence that they had to be 'developed' and 'civilised' (Jazeel 2012). By 1914, Europe held dominion over more than 85% of the globe (Morrell & Swart 2005, p. 61). For Marx, colonialism was a system of exploitation that required not merely the formal annexation of countries but their *economic subordination* to the reproduction of capital in the dominant Western countries (Pradella 2013, p. 122). The subsequent underdevelopment of colonies and regions peripheral to the West/Global North has been central to processes of capitalist development, with former colonies and developing countries continuing to provide access to cheap raw materials (Ghosh 2019).

Central to Moore's (2017) apt diagnosis is a Marxist critique of the socio-metabolic reproduction of capital and its need to continuously exploit people as well as the natural world in its boundless quest for new markets in order to maximise shareholder profits. These dynamics are further exacerbated by capital's iteration of the Cartesian separation (see Chapter 2) of humans from nature (Plumwood 2002) and producers from the means of production (Soriano 2018). Since the advent and spread of capitalism, a profound alienation is seen as increasingly corroding interpersonal relations, wherein people (human and nonhuman), places and things appear as mere commodities to be traded for a given exchange value. Other-than-human animals in particular, traditionally rendered mute objects devoid of agency and subjectivity, are further reduced to mere living material for biotechnological agriculture, cosmetic and pharmaceutical industries, and related profit-seeking enterprises (Berger 2009; Braidotti 2009, p. 529). Hence, utopian scholar Darko Suvin asserts:

> Capitalism has by now grown fully parasitic ... the system survives only by continuously increased extortion of surplus labour from the 95 percent of lower classes and nations to the rulers. It is by far the biggest manmade threat to liberty, cognizing, and creativity invented by our species. Or simply to survival. (2021, p. 24)

Capitalism is a system that increasingly takes, leaving little but ruin in the form of socioeconomic precarity for the multitudes and ecological devastation in its wake. Its unprecedented 'pitilessness' is one that increasingly spares no one (Berger 2007, p. 87). Thus, following Suvin and others, Marxist feminist Nancy Fraser refers to capitalism – especially in its present globalising, neoliberal[4] and financialised form – as the 'underlying object of our general crisis' … a predatory and ultimately unstable system that undermines the political, ecological, social and moral conditions of its own existence (2019, pp. 37–38). The grossly unequal power relations, and vast geographical, cultural, structural, socioeconomic and historical imbalances that continue to underlie contemporary access to biophysical resources and related impacts on the biosphere challenge the 'assumption of humanity as a homogenous driver' (Pichler et al. 2017, p. 32). For instance, the explosive proliferation of human economic activity that marked the 'Great Acceleration' period denoted earlier overwhelmingly took place in the Global North (Pichler et al. 2017). The latter stages of the fossil economy – electricity, the internal combustion engine, the petroleum complex – were effected through investment decisions by owners of commodity production with occasional involvement by certain governments, not through widespread democratic deliberation and consent (p. 64). Significant swaths of the world's population still remain excluded from the fossil economy, even lacking access to electricity (Mal & Hornborg 2014, p. 65). Meanwhile, the planet's top emitters – the US, EU, and China – account for nearly half of global CO_2 emissions (Piketty & Chancel 2015). Moreover, the notable Carbon Majors Report published in 2017 revealed that the top 100 fossil fuel companies accounted for more than 70% of emissions since 1988 (CDP 2017). Recent analysis by the Climate Accountability Institute found that just twenty companies – including the usual suspects Shell, BP, Exxon and the even more polluting state-owned Saudi Aramco – have contributed over one-third of all energy-related carbon and methane emissions globally since 1965. As Mal & Hornborg rightly observe, 'Depending on the circumstances in which a specimen of Homo sapiens is born, then, her imprint on

---

4   Here I draw on Harvey's (2007) list of key distinguishing characteristics including deregulation, accumulation by dispossession, privatisation and an upward redistribution of wealth, which in turn have led to a number of contemporary perturbations – i.e. socioeconomic insecurity, cultural anxieties, political polarisation, ecological breakdown.

the atmosphere may vary by a factor of more than 1000' (Mal & Hornborg 2014, p. 65). Indeed, the average lifestyle consumption carbon footprint of the richest 1% is estimated to be more than seventy-five times that of the poorest 50%, at the conservative end (Bruckner et al. 2022). What's more, in a cruel twist of irony, developing nations and marginalised populations the world over, who have contributed the least to contemporary climate and ecological breakdown, are already suffering the worst of the impacts.

However, Amitav Ghosh (2021) does well to warn of the deceptive allure of the 'Fog of numbers' – i.e. the focus on per capita footprint in mainstream climate change debates. This limited framing can create a highly skewed picture of the major drivers of climate breakdown by over-emphasising the role of 'individual responsibility' and thus occluding the significant impacts of '*institutional emissions*, like those related to the US military and to the projection of American power' (p. 151). Thus, both hyper-individualistic emphases and the abstract attribution of species-level blame for contemporary socio-ecological decline are misguided. Similarly problematic are the resurgent neo-Malthusian diagnostic framing narratives denouncing the 'measureless multiplication of (racialised) human beings' (Lorenz 1973, p. 11). The 'limits-to-growth' discourses which emerged during the 1960s – as evinced by landmark works in this vein such as Paul Ehrlich's *The Population Bomb* ([1968] 2009), Meadows et al.'s *The Limits to Growth* (1972), and more recently Clarke and Haraway's *Making Kin not Population: Reconceiving Generations* (2018b) – draw attention to the human species' looming breach of the earth's carrying capacity through economic and population expansion. However, focusing on numbers alone or even as a major driver of climate and biodiversity breakdown serves as a red herring that distracts from their structural origins. As discussed in the preceding section, a growing body of empirical evidence clearly demonstrates that it is certain powerful actors – the Global North, multinational corporations and the profligate consumption of the wealthy elite – who exert astronomically greater impacts on natural systems and other species than the world's poorest (Otto et al. 2019; Dow & Lamoreaux 2020). This alone shows that the problem lies less with sheer numbers than with particular ways of living and being which have become hegemonic in Western-capitalist societies – mass conspicuous consumption patterns that appropriate disproportionate amounts of resources and energy, thus exerting impacts that dwarf those of scores of denizens in developing countries combined. Of course, all CO_2 molecules effect the same radiative forcing – i.e. atmospheric warming effect. However, there lies a profound ethical gulf between what the philosopher Henry Shue (1993, 2019) termed 'luxury' versus 'subsistence' emissions. The former originates from the profligate status consumption on behalf of the world's wealthiest – SUVs, private jets, super-yachts, sprawling mansions; the latter stem from processes of physical reproduction by the poorest, who often have no choice but to use energy from coal-fired power plants, for instance, in the business of survival itself.

As human ecologist and activist Andreas Malm (2021) poignantly observes:

> Subsistence emissions must be overcome just as much as any other, but they have none of these features of luxury in a CO2-saturated world: wanton criminality, insulation from the fallout, waste promotion, withholding of resources for adaptation, persisting in the most odious variants and ostentatiously negating the very notion of cuts. (92)

Hence, recent calls for the need to shift focus/policies towards high-emitting countries & restricting consumption patterns of the super-rich (Otto et al. 2019; Bruckner et al. 2022). Indeed, the working group III contribution to the IPCC's Sixth Assessment Report (2022) has at last begun to explicitly acknowledge the ecological impacts of such gross disparities, noting that 'the top 10% of households with the highest per capita emissions contribute a disproportionately large share of global household GHG emissions' (p. 8). Thus, the more precise concept of 'Capitalocene' brings a notably critical refinement to mainstream 'Anthropocene' discourses by shedding crucial light on the socio-ecological depredations of a particular set of exploitative socioeconomic and historical patterns, rather than designating an undifferentiated 'humanity' as the driver of global ecological decline (Moore 2017, p. 597). Yet capitalism is centreless, an assemblage of innumerable actants variedly entangled in ever-shifting webs of micro and macro-interactions. This makes it difficult to designate a single individual or even a group of actors as wholly culpable for its socio-ecological disturbances. Neither is there a 'collective subject' that we can attribute blame to (Fisher 2009, p. 66). However, if we conceive of culpability as a spectrum with degrees of wrong-doing and responsibility, powerful actors such as wealthy elites and transnational corporations who wield their enormous wealth and political power to influence institutions, media platforms, political processes and public sentiment in their favour undoubtedly emerge as prime offenders. Nevertheless, to focus on the super-wealthy and the socio-ecological injustices of their excesses as aberrations of some sort would be a mistake – they are symptoms of *systemic ills* that must be uprooted entirely so as to create a world wherein neither those in want of homes nor billionaires exist.

Brave New Worlds

Some such as environmental scientist Erle Ellis (2011) have sought to frame the 'Anthropocene' as a potentially positive development offering novel opportunities for channelling or doubling down on human resilience and ingenuity for moulding the new human planet into a prosperous new era. Central to the 'good Anthropocene' narrative is an unchallenged resolute faith in the purportedly boundless potentiality of human technological prowess, which facilitates the control and prediction of the natural world, and can help humans overcome most if not all ecological woes or barriers that might emerge (Hamilton 2015). Ellis muses thus:

> Humans have dramatically altered natural systems – converting forests to farmlands, damming rivers, driving some species to extinction and domesticating others, altering the nitrogen and carbon cycles, and warming the globe – and yet the Earth has become more productive and more capable of supporting the *human* population. [emphasis added] (p. 38)

Ellis urges humans to harness their 'unprecedented and growing powers' (p. 38) and to honour their responsibilities as stewards of this brave new world. The first point to make is that the myriad crises we're experiencing stemming from fundamental alterations of earth systems demonstrate precisely the opposite of Ellis's assessment. The Anthropocene denotes a rupture in human-earth relations, hence its reference to a new geological epoch, not an acceleration in human mastery over ecological change (Hamilton 2015). Moreover, such a position draws on the anthropocentric hubris (Washington et al. 2021) that in many ways led to our current predicament, reinforcing the belief about humans as the hegemonic species and suggesting that this is our world to do with as we wish. As the philosopher Clive Hamilton (2015) observes, 'In the service of the inherent dynamic of Progress the negativity of ecological damage is sublated or assimilated into a positive force for change' (p. 235). The golden promise of a new dawn ushered in human technological ingenuity, continues Hamilton, 'lulls them [the victims of the current crises] into silent endurance' (2015, p. 236). It is never too late to act to reduce and prevent more suffering. The widespread loss of life and disintegration of natural systems should be resisted, and wherever it's too late, we ought to work collectively to (re)create habitable dwelling spaces wherein all terrestrials might flourish anew. Relatedly, mainstream 'Anthropocene' discourses often fail to challenge the centrality of human agency implied in the concept, framing humans as the primary Earth movers. In a mere handful of generations, industrial capitalism has virtually exhausted fossil fuel sources that took hundreds of millions of years to accumulate (Crutzen 2002). Thus, this era is declared the age of an all-powerful and ubiquitous human agency. On the other hand, as we have already seen, one of the great paradoxes of the Anthropocene is that we are witnessing powerful other-than-human agencies in the form of viruses, hurricanes and the like take centre stage. Crucially, as I will try to demonstrate via examining myriad manifestations of ecotopianism in the coming chapters, the historical, cultural, and socioeconomic trajectories that have led to the Anthropocene were by no means inevitable. Industrial capitalism is not predetermined but particular a socio-historical contingency bolstered by unusually favourable environmental conditions throughout the Holocene that paved the way for our industrial-agricultural ways of life (Chakrabarty 2009). Thus, our present predicament *can be otherwise*; more equitable, sustainable, post-capitalist ways of life conducive to interspecies flourishing *can* be brought forth. Because ultimately, oppressive systems like capitalism would be nothing without our cooperation and the billions of micro and macro-interactions and alliances that keep them afloat (Fisher 2009, p. 15).

Navigating Interspecies Imbroglios in and Beyond the Capitalocene

> ... If anything, the Anthropocene is a spark that will light a fire in our imaginaries. This is a time to think big, to dream. We dream about abundant futures.
>
> – Collard et al. (2015, p. 326)
>
> What if? That humble question might be the single spark that can ignite a prairie fire, provoking us to leap beyond personal speculation and into the vortex of political struggle and social action. *This is how it's always been; this is the world as we've always known it.* But why is it so? Who benefits and who suffers? How did we get here and where do we want to go?
>
> – Ayers (2016, pp. 7, 8)

Complete stability and harmony have never been features of our dynamic earth system, which has been beset by all manner of historical cataclysms that have paved the way for new evolutionary potentialities. However, what is clear is that colonial-capitalism's impacts on the environment and on other species have reached 'omnicidal' dimensions (Ghosh 2021). A growing chorus of voices bemoan the mounting, widespread and irrevocable losses – of baselines from which to orient ourselves and assess the damage, of socio-ecological integrity, and of our terrestrial counterparts. Often exhortations are made for the urgent existential imperative to intervene in contemporary ecological disintegration in order save the Earth *from* ourselves and largely *for* ourselves. Paradoxically, homo sapiens are simultaneously presented as the dominant global Earth movers, whose extensive reach is both disavowed and yet called upon to help stem further decline, alongside recognitions of humanity's utter powerlessness in the face of increasingly unpredictable socio-ecological perturbations. Following Haraway (2015), perhaps it might be more useful to think of the Anthropocene/Capitalocene as a boundary event rather than an epoch, marking severe spatial-temporal discontinuities wherein what comes next, however uncertain, will likely not be like what was before (160). The new terrain in and ahead of this boundary event is characterised by material reality's 'vengeful reassertion of itself' (Žižek 2011, p. 330), which has in turn generated a philosophical earthquake that has 'unsettled the tectonic plates of conceptual convention' (Johnson et al. 2014, p. 447). The arrival of the Anthropocene has signalled the 'public death' of longstanding humanist presumptions of Nature as separate from a privileged Social realm (Beck 2010; Latour 2018), of 'natural' history as distinct from 'human' history (Chakrabarty 2009), and of 'Nature' as a mythical construct designating an Edenic state of pristine natural harmony preceding the arrival of humans (Cronon 1996; McKibben 2006; Bennett 2010; Morton 2010; Latour 2018). Now we are increasingly presented with dynamic nature-culture entanglements, wherein we must contend with an array of other actors and an animate earth that 'vibrates underfoot' and reacts, sometimes violently, to our actions (Latour 2017, p. 149).

Though we cannot return to a mythical state of ecological harmony devoid of human presence, we must reckon with past 'discursive-material processes of annihilation, displacement, and replacement' driven by imperial capitalism that have led to our current state of 'ruination and ecological impoverishment' (Collard et al. 2015, p. 323). Crucial queries for forging ahead beyond the Capitalocene, and for any ecotopian to keep in mind, include: how do we respond to and navigate the hybrid human-animal-nature entanglements in ways that are far less destructive and far more conducive to multispecies flourishing? How do we build worlds 'literally filled to the brim with different creatures' with whom we share our earthly and mortal existence (Latour 2004, 2014), not only for our sake but, crucially, for theirs as well (Collard et al. 2015, p. 323)? In enquiring about how to build more ethical and liveable futures, I deviate from proposals like those put forward by philosopher Patricia MacCormack in *The Ahuman Manifesto* (2020), which calls for 'an end to the human both conceptually as exceptionalised and actually as a species' (p. 5). The first part of MacCormack's proposition, which seeks 'the end of the anthropocentric world – ultimately the end of humans' violent occupation of Earth' (p. 7), is one that I sympathise with, as well as the need to deconstruct the 'human' as an 'exceptionalised' species. Such critiques of 'Anthropocentrism' as a cultural dominant that oppresses, exploits, vilifies and reduces other species to the inferior status of object, as we will see in the following chapter, are central diagnostic narratives within many green utopian manifestations. However, MacCormack's assertion that an end to 'humanity's' violent occupation of the earth requires, among other things, our non-existence as a species (through a cessation in human reproduction) so that other species might thrive – is deeply problematic. For one thing, it is politically fatalistic, and it commits a similar error as universalising 'Anthropocene' narratives by making the essentialist suggestion that *humans* are the problem, rather than particular modes of subsistence (i.e. industrial-colonial-capitalist) and relationality (i.e. Anthropocentrism). As I'll expand upon in Chapter 6, myriad indigenous cultures historically and presently serve as a testament to the possibility of humans living more respectfully and ethically with other species and terra. As Haraway (2015) points out in her designation of the present time as the Chthulucene in reference to the ongoing forces and powers spanning the past, present and to-come within which all terrans are enmeshed, charting viable pathways will require concerted commitment on our part to collaborating with our co-terrestrials. The task, in other words, ought to be to learn to make kin with human and other terrestrials – in effect to radically rethink the collective 'we' – as we engage in the vital work of recreating new lifeways.

The aim of this chapter was to provide an overview of the singular – and paradoxical – era we presently find ourselves in, one laden with intensifying socio-ecological upheavals, and which serves as the unique historical context inspiring the emergence of radical alternatives. This book aims to further elucidate the complexities, trials and tribulations of learning to live well and 'make kin' (Haraway 2016) in the Capitalocene from the perspective of contemporary manifestations of ecotopianism – from canonical ecotopian literary texts to REAs putting their bodies on the line in order to physically stem further tides of loss. Green utopianism in its myriad manifestations holds immense value for resisting and building beyond our troubled present by offering trenchant critiques of the socio-ecological status quo as well as imaginative projections of more ethical, equitable and inclusive alternatives. Subsequent chapters will feature in-depth examinations of contemporary green utopian forms attempting to (1) challenge the 'cultural dominants' of Western Anthropocentrism and contemporary capitalism and (2) radically reimagine how we might relate differently to our co-evolutionary kin. However, as outlined at the beginning of this introductory chapter, of specific interest will be: to what degree do these various green utopian modalities challenge Anthropocentric views of human separateness and superiority to the natural world? How are other species framed – as inert resources for human use, or agentic kin whom we ought to respect and learn to live well with? Who is included (and excluded) in the collective 'we' constituting the *demos* of a better society – all sentient animals? All living, or even non-living components of the biosphere (Bennett 2010)? What are the contrasts and continuities between ecotopian social movements and literary works in terms of how they engage with these important ethico-political questions? Before we proceed to analysing and discussing particular ecotopian forms, the next chapter will offer a brief and critical overview of the history and key tenets of Western Anthropocentrism, followed by a critical discussion of the two theoretical tributaries informing this book: (green) utopian and critical-posthuman theory (Chapter 3). The latter, it is argued, serve as invaluable tools for assessing the limitations as well as the potential of varying visions of better worlds, and most importantly, the role of our long-excluded terrestrial kin in them. It is high time, not to learn to die, but to learn to fight in the Capitalocene for life and for better worlds.

CHAPTER 2

Supremacy

This chapter will feature a brief critical discussion of the history, key features and lingering legacies of Anthropocentrism, a Western cultural dominant that positions humans as apart from and above other species in a great hierarchy of being. For many ecotopian manifestations, as we'll explore in the following chapters, Anthropocentrism is designated as a key underlying driver of contemporary socio-ecological decline. Western anthropocentrism and the entire constellation of attitudes, values, worldviews, behaviours, rights and privileges associated with this regime has a long and complex history with both secular and religious tributaries that has been discussed at length elsewhere (Midgley 1983; Plumwood 2002; Steiner 2010; Wolloch 2017). Here I will focus on a few key developments and thinkers who have stood out to me throughout my years of research for their landmark work in helping to deconstruct human supremacy. Moreover, though I repeatedly make reference to 'the West' and 'Western culture' as being particularly Anthropocentric, this is not to say that there haven't been radical counter-currents within Western traditions that have challenged these ideas. The deep ecology movement and contemporary REAs (Chapter 4), and the ecological teachings of Saint Francis of Assisi which ascribed personhood to the sun, other animals and plants, to name a few, emerge among countless others who have fervently espoused and advocated for alternatives to the central tenets of Western humanism. What I try to draw attention to in this chapter are dominant traditions shaping human relations with other species prevalent especially in Western cultures, in order to better highlight one of the oppressive systems that green utopianism seeks to transgress and move beyond. Lastly, when discussing human supremacy one should never lose sight of the 'pervasive racialized hierarchies within "the human" [which] result in the discounting of millions of non-Western, non-white, non-affluent, non-adapted, and non-resilient individuals as "less-than-human"' (Butler 2004; Marhia 2013; Tschakert et al. 2021). Speciesism stems from similar structures of oppression and exclusion which lead to racism, misogyny, homophobia, transphobia and ethnophobia. Exploitative systems such colonialism, capitalism, patriarchy and Anthropocentrism have been 'ending worlds for as long as they have been in existence' (Yusoff 2018, p. 11). They recall Robert Oppenheimer's fateful recitation of the famed quote from sacred Hindu text the Bhagavad-Gita, 'Now I am become death, the destroyer of worlds', upon witnessing the detonation of the world's first nuclear bomb in 1945 (Temperton 2017). This is especially true of the horrors variedly inflicted on our terrestrial kin in the wake of Western capitalist humanity's ongoing appropriation of the earth for itself alone.

Aloof and Aloft: The Legacy of Western Humanism

Kemetic (ancient Egyptian) and Confucian traditions tend to view the human 'self' as a field of relationships, an individual-in-relationship or person-in-community (Yin 2018, p. 204). In stark contrast, a central pillar of Western thought rests on the assumption that individuals – seen as distinct and separate from one another – exist *prior* to social relations (Yin 2018). This is important because it informs the related cultural dominant of Anthropocentrism, the belief in human 'subjects' as categorically separate from and superior to an inert 'nonhuman' world (Plumwood 2002; Latour 2004; Ferrando 2016). Anthropocentrism as a system of human supremacy and domination over other species, founded on a declaration of absolute human sovereignty, emerges as a key feature of Western modernity (Steiner 2010). Anthropocentrism, or 'human exceptionalism', in the words of philosopher Vinciane Despret (2022), is the 'sordid habit of placing the human at the centre of the world and of its stories' (p. 20). As philosopher Jacques Derrida poignantly observes in his posthumous deconstruction of the human/animal binary *The Beast and the Sovereign* (2010), affirmations of sovereignty serve as a 'performative, a commitment, an act of sworn faith, of *war declared against a sworn enemy* (i.e. other species): sovereignty is a posited law, a thesis or a prosthesis, and not a natural given ...' (Ernst Friesenhahn in Derrida, p. 77). To declare sovereignty (in an absolute sense) is in effect to sever equal relations based on kinship or fellowship by establishing an external 'other'; this in turn initiates a great conceptual antagonism between humans and other species in the West. The human sovereign, as the self-declared purveyor of laws, of all benchmarks of significance, declares themselves entitled to the unbridled use of other species. Worldviews and attitudes matter; ontology – beliefs and understandings about the nature of 'reality' – influence ethics, or how we feel we ought to relate towards others. When you see another as kin with whom you are intimately and inextricably entangled – a central tenet of many Indigenous ethical and cosmological systems, as we will see in Chapter 6 – different relations follow, typically based on respect and compassion. That which is not a fellow and is considered external to you can therefore more easily be classed as 'there for the taking', to be used and exploited at will. From this stance of radical separation, it becomes easier to discount their wellbeing and dismiss them as inconsequential. By elevating and privileging the human above all else, Anthropocentrism subordinates all others, casting them aside as ethically and ontologically inferior.

Anthropocentrism, like Western thought more generally, is pervaded by dualisms and hierarchies. Historian Lynn White in his famous and somewhat controversial essay *The Historical Roots of our Ecological Crisis* (1967) argues that many of today's ecological woes have their roots in Judeo-Christian dominion theology, particularly the Old Testament. Well-known passages in this text refer to a firm hierarchy ordained by God, wherein humans – as the pinnacle of creation – are urged to exercise dominion over the Earth and its nonhuman inhabitants.[1] Of course, there are many tributaries within Western culture and theology which challenge anthropocentric visions of superiority and dominion. White alludes to the example St

---

1   Indeed, cross-national studies like that by Schultz et al. (2000) have found that those who express literal beliefs in the bible tend to exhibit more anthropocentric concerns over the environment (i.e. one should act on climate change because it threatens *humans*).

Francis of Assissi, an Italian friar (1182–1226) who was known for cherishing other species as fellow creatures of God and referring to them in an egalitarian manner as 'brothers and sisters' (Crane 2021). Assissi is one figure who exemplifies what utopian scholar and activist Tom Moylan denotes as the utopian potential of religious discourse (2021, p. 69). Nevertheless, the influential Greek philosopher Aristotle's (384–322 BC) Escala Naturae, or Taxonomic scale of nature, arranged different beings at different levels of perfection (i.e. mammals over reptiles). In *Politics* (350 BCE) Aristotle famously mused that, though human beings are animals, all other animals exist for the sake of 'Man' (p. 13). Here the gendered term is noteworthy, as Aristotle saw women as biologically and intellectually inferior to men. The late environmental philosopher and eco-feminist Val Plumwood in her landmark work *Feminism and the Mastery of Nature* (2002) demonstrated how the logic of hierarchy and dualism works to segregate a whole host of historical 'others', paving the way for their unbridled exploitation. This legacy can also be traced to philosopher René Descartes, sometimes referred to as the 'father of modern philosophy' (Armstrong 2003, p. 128). Descartes theorised a mind/body dualism which initiated a dominant rationalist legacy that privileged the disembodied and incorporeal *rational mind* (which was deemed to constitute one's entire soul) over the material and mechanistic body of the organic world. Humans alone, specifically men, were sites of a disembodied and autonomous consciousness, which conferred free will and self-determination. Other species, on the other hand, were deemed to lack such prized abilities and were therefore associated with the inferior 'mechanistic' body of the natural world. Indeed, for Descartes, other animals were little more than 'machines ... made by the hands of God' (1637, p. 41). An array of other dualisms can be traced alongside this consequential schism – Culture/nature, Human/animal, Male/female, West/non-West. For Plumwood (2002), dualism constitutes a system of domination within hierarchical relationships between a 'Master Identity' – i.e. white, heterosexual, European men – and its 'others' – i.e. women, nature, other species, other-than-Western peoples. Crucially, power shapes the identities of both sides of the hyper-separated, contrasted as 'Higher' vs. 'lower', 'Center' vs. 'periphery', 'Active' vs. 'passive' (Irving & Helin 2018). These dualistic relations are further marked by certain key features, including: radical exclusion, wherein the 'other' is not only seen as different but completely separate from and inferior to a given 'Master identity'. The Western logic of dualism and denial of interdependence – in casting the 'other' as separate, inferior and therefore sacrificial – has served as the theoretical justification for their exploitation and oppression within colonial, patriarchal and capitalist systems (Plumwood 2002; Klein 2020).

Anthropocentrism thus frames humans (though not all in the same way) as the ultimate 'Master Identity' – the prime locus of value and significance, agency and intentionality. Other species, our most radically separate and inferior 'other', have been variedly portrayed as soulless, mindless automata, reduced to the status of *unseeing* objects merely *seen* by knowing human subjects (Derrida & Wills 2002, p. 383). The language and systems of categorisation we use also reflect such hierarchies and dualisms (Derrida 2002). For instance, we tend to relegate other species to the homogenous category of 'animal', with humans set apart and aloft, thereby both denying our own animality and violently reducing their innumerable singularity. Indeed, their reduction and simplification serve the interests of anthropogenic power (Berger 2007, p. 134), just as the denial – and homogenisation – of other-than-Western cultural diversity facilitated colonial ambitions of domination (Krenak 2023). One could proceed with 'human' and 'nonhuman animal', which at least correctly implies that humans too are animals. But this too reproduces a dualistic ordering, wherein once again all 'nonhumans' are lumped together in one homogenous category and, crucially, defined as 'lack' (non) in relation to the human 'Master Identity'. As such, Donna Haraway (2008) draws on queer theory with its essential job of 'undoing normal categories' to propose an alternative term, 'companion species' for designating us terrans 'in all our off-category kin and kind ... the stuff of being in material semiotic intra-action with no ontological starting point' (p. xxiv), no 'us' as categorically separate from 'them'. Expanding on this, I propose the term 'terrestrial kin' which I will draw on throughout, in order to even more strongly underscore the myriad ways in which we are intimately entangled in the lives and fates of Earth others (even though this implies the exclusion of extra-terrestrials!) across vast spatial-temporalities. Terrestrial kin also, in my view, in underscoring our earthbound, co-evolutionary affinity with other beings is particularly transgressive of hierarchies or antagonistic dualisms. Earth others are in an evolutionary and companion sense our kin, persons with whom we are linked in terms of heritage, and by virtue of our shared mortality and earthly habitation. Viewing another as kin means that treating them as expendable objects or tradeable commodities with mere instrumental value becomes exceedingly difficult, if not impossible.

Spheres of Significance

Attitudes of separateness and superiority inherent within human supremacy result in the exclusion of the other-than-human, or at least certain kinds of species, from consideration as ethical and political subjects and equal members of a wider collective. Immanuel Kant's famed cosmopolitan proposal in *Toward Perpetual Peace* (2005) for a global citizenry bound by universal law and solidarity, in addition to being decidedly Western-Euro-centric, did not include beings external to the human-world correlate. Rather, Kant regarded only *rational* beings as worthy of being treated as ends in themselves; non-rational beings (i.e. other species) had only 'relative worth, as means, and are therefore called *things*' (Kant & Schneewind 2002, p. 496). Similarly, the influential political philosopher John Locke muses thus in landmark text *Second Treatise of Government* (2000): 'there cannot be supposed any such subordination among us, that may authorize us to destroy one another, as if we were made for one another's uses, as the inferior ranks of creatures are for us ...' (Ch. 2, p. 6). Certain characteristics such as speech and humans' command of the written word, closely associated with reason and therefore 'humanity', worked to further exclude other species from the community of moral and political subjects. As a result of these dominant legacies, contemporary political thought on the whole continues to construe the 'cosmos' or collective 'we' in decidedly narrow, anthropocentric terms, wherein humans feature as the only entities on the negotiating table (Latour 2004a, p. 454; Stengers 2010). Similarly, within the tradition of Roman law, other species have traditionally been regarded not as persons but as property – i.e. consider common references to dogs as beings we 'own' rather than live with. Although promising developments are underway in this regard – for instance, since 2000 across numerous US cities the legal status of certain other-than-human animals has changed from 'property' to 'companion' (Balcombe 2016). Other landmark recent developments include the US-based Nonhuman Rights Project, an organisation which fights to secure the legal personhood and rights of other-than-human clients such as elephants, apes and dolphins, and the new UK Animal Welfare (Sentience) Bill which will recognise vertebrate animals as sentient – i.e. the capacity to be aware of oneself, one's interactions with the surrounding environment and other beings, and to experience pleasurable (happiness) as well as aversive states (pain, fear, etc.) (Broom 2007, p. 100) – in domestic law. Recently, following a comprehensive LSE report (Birch et al. 2021) which concluded that decapods (i.e. crabs, lobsters) and cephalopods (i.e. octopuses) are indeed sentient, the bill has been amended to recognise them as such. It cited that, unlike other invertebrates, decapods and cephalopods boast 'complex nervous systems, one of the hallmarks of sentience' (DEFRA et al. 2021). As such, the traditional practice of boiling a lobster alive could soon become a criminal offence. Similarly, the food corporation Nueva Pescanova's plans to start farming octopus has been heavily criticised recently due to ethical concerns over the mass production of these highly sentient creatures (Hamilton 2023).

The recategorisation of decapods and cephalopods as beings who matter ethically and legally undoubtedly marks a step in the right direction by reducing the unnecessary suffering of groups of animals who have been long-excluded from ethical considerability. However, it also has serious limitations: by focusing on sentience, species deemed to lack this characteristic risk being excluded from consideration as persons with rights. Sentience has long served as the bedrock of traditional debates around animal ethics (Singer 1973). The idea stems from English philosopher Jeremy Bentham's famed dictum, 'The question is not, Can they reason? nor, can they talk? But, can they suffer?' (1789). This is in direct reference to the longstanding anthropocentric assumption that other animals don't matter, or matter less than humans, because they purportedly lack 'human' characteristics such as rationality and language. Animal rights philosopher

Tom Regan ([1986] 2004) emphasised animals' advanced cognition and their capacity to be self-aware as essential foundations for their moral consideration and (re)categorisation as rights-bearing individuals. For Regan, crucially, what matters is whether a particular organism/species is the 'subject of a life'. However, this definition prioritises vague (and once again human-centred) criteria such as the ability to have beliefs, a sense of the future, memories, and emotional lives in order for a particular being to count as an ethical subject (2004, p. 187). In effect, ethical valuation on the basis of arbitrary characteristics such as sentience and intelligence functions as a conditional mode of justice wherein beings are granted rights or consideration on the basis of their possession of 'morally significant characteristics' (Wolfe 1998, p. 13). Characteristics believed to be exclusively possessed by humans, namely the capacity for language[2] and complex thought, have only gradually and begrudgingly been granted to a select handful of species (i.e. cetaceans, primates, corvids) by humans as the ultimate arbiters of such matters.

The recent 'compassionate conservation' movement within conservation biology, though laudable for its emphasis on reducing intentional harm to other species and promoting more peaceful coexistence, also retains a focus on sentience as the moral benchmark (Wallach et al. 2020; Coghlan & Cardilini 2022). Relatedly, Lori Gruen's (2015) ethical concept of 'entangled empathy' rightly calls for attending to the wellbeing of other animals with whom we're always 'entangled' in an ongoing effort to improve our relations with them. However, for Gruen, empathic entanglement and engagement can only occur with beings whose life-worlds we can imagine, a boundary that she draws at vertebrate life. Plumwood (1991, 1993, p. 193; 2002), on the other hand, seeks to deconstruct the inert/agentic divide by advocating a philosophical animism that sees both the sentient and more-than-sentient world as a wellspring of agency and intentionality, with whom we can and should attempt to cultivate mutual empathic interchanges despite the difficulties laden therein. To do so

2   The notion that humans have a monopoly on speech or communication is yet another expression of anthropocentric arrogance: 'In fact, the Earth speaks to us in terms of forces, bonds, and interactions, and that's enough to make a contract' (Serres 1995, p. 39).

is not necessarily to effect a 'narcissistic projection' of our own interests and desires onto the 'other' (Gruen 2009, p. 32). Moreover, this does not mean that it won't be easier to engage with the lifeways of species with whom we share closer evolutionary and experiential affinities, like other mammals. However, to limit attempts at empathic engagement and ethical relationality to vertebrates – or to 'sentient' beings for that matter – functions as an ethical double-edged sword that enhances empathy for some (sentient) species whilst excluding a multiplicity of other forms deemed to fall short of such prized attributes (Plumwood 2002). Indeed, the UK Sentience Bill mentioned above excludes invertebrates. Thus, what of the fate of increasingly threatened entomofauna (Sánchez-Bayo & Wyckhuys 2019) – i.e. beetles, wasps, bees, ants – perceived to be so unlike us as to be rendered virtually invisible?

The aforementioned conditional modes of justice constitute examples of ethics based on affection for sameness (Levinas 1989; Sargisson 2000). Anthropocentric orientations make it so that we value others on the basis of their likeness to or usefulness to us. Whales, for instance, have captured the public imagination across the Western world largely because decades of research have revealed that they lead highly complex emotional and social lives, like us. Yet paradoxically, other animals and the numerous ways in which they are unlike us have also been central to how we define ourselves; as we have seen, humans are (so the logic goes) *not* like other animals because supposedly we alone can speak and reason; *we* can formulate abstract concepts, and have a sense of our own mortality; *they* lack such abilities. Of course, contrary to long-established orthodoxy (Rolston 2015, p. 39) humans are far from the only reflective or moral species. Nevertheless, such characteristics are variedly utilised as justifications for the perceived chasm between humans and other species. As John Berger observes, other animals' supposed silence 'guarantees its distance, its distinctness, its exclusion, from and of man' (2009, p. 14). So, we have long defined ourselves in relation to what they seem to lack, a phenomenon that political scientist John Rodman (1980) terms the 'differential imperative'; nonhumans mark and reinforce the bounds of our humanity. That which is human is continually reified and delineated in juxtaposition to nonhuman 'otherness'. Similarly, metaphors and constructs of monstrosity throughout history and across cultures have often been predicated on and variedly alluded to 'nonhuman' otherness (Moser & Zelaya 2020). The monstrous, decidedly un/inhuman, is deemed more akin to animality. Moreover, around the world (Chow et al. 2022) struggles continue to rage over which humans count as 'human', wherein 'Proximity to animals can be deadly – even if this closeness is just the result of a visual juxtaposition or a racial slur' (p. 9). Many BIPOC and similarly marginalised peoples 'continue to bear the burden of imposed comparisons to the nonhuman' (Chow et al. 2022, p. 9).

Western humanity has arrogated to itself the exclusive position of the *seeing* subject (Derrida & Wills 2002, p. 383). That other species could regard us from their own vantage points, with their own valid ways of knowing and being in the world, is a notion that has variedly been dismissed as naive, and wildly improbable if not impossible. Indeed, the assumption of full knowledge of the animal other is itself 'an index of our power, and thus an index of what separates us from them' (Berger 2009, p. 27). We alone are all-knowing. Other terrestrials thus appear worlds away – both ontologically and epistemologically inferior, 'poor in world' in Heidegger's infamous estimation (Elden 2006). However, as Rose (2013) denotes, consider the 'concealed attentiveness' of the crocodile which, like any good hunter, 'knows others are paying attention, they know the ways in which others pay attention, and they find ways to circumvent that attention' (p. 103). Moreover, anyone who has lived with companion species such as dogs or cats will likely wonder how anyone could have ever come to such erroneous conclusions, knowing full well the array of unique personalities and aptitudes displayed by each individual. My dog Luna recognises subtle tonal shifts, and the meaning of a range of words and phrases in both Spanish and English. She is an exceptionally keen observer of human body language, and knows precisely what the myriad different gestures and mannerisms enacted by my family and I signify. Having keenly observed us every day for the better part of a decade, she has come to understand that the gate before our front door opens inward, and therefore that she must use her paw to pull rather than push it if she wishes to go outside.

> Oxen that rattle the yoke and chain or halt in the leafy shade, what is that you express in your eyes? It seems to me more than all the print I have read in my life. (Whitman in Muldoon 1997, p. 238)

We will likely never know the answer to such queries, and it's just as well, for 'valuing life is so much more important than simple comprehension' (Gumbs 2021, p. 39); the latter will forever be eluded by the inexhaustible mysteries and wonders of the other-than-human world. But we can aim towards a relative mutual understanding and respectful coexistence. Yet this goal is significantly hampered by our consignment of other species to a realm of inconsequentiality, wherein they are often seen as little more than resources for the satisfaction of *human* needs and desires, rather than as affective persons who matter in their own right, with needs and desires that ought to be respected. Worse, and unsurprisingly, Anthropocentric attitudes and value systems have provided the necessary foundations for the systematic persecution, enslavement, brutalisation, slaughter and dispossession of other terrestrials for thousands of years. We've razed their forest homes, polluted and dammed rivers, made the oceans more acidic with industrial activity and anthropogenic climate change, and otherwise stolen their livelihoods and very futures. Some estimates suggest that up to 70 billion land-based animals are killed every year (Sanders 2018) in the most horrific ways as fodder for the agricultural-industrial complex, whilst trillions of wild-caught fish meet the same grim fate every year. We test a range of products on them as a matter of course, from cosmetics to pharmaceuticals, often causing unimaginable distress. Or at best their plight is ignored, as with the countless news headlines about humans killed by extreme weather events while the suffering and loss experienced by other species is rarely acknowledged. Hence, perhaps the most unsettling and dangerous effect of Anthropocentrism is that it has rendered the current sixth mass extinction all but invisible in the eyes of the public and mainstream media (Crist & Kopnina 2014). Climate breakdown has increasingly, and understandably, captured the public imagination for a host of complex reasons, namely greater media coverage and a growing recognition of it as an existential threat to life on earth. However, though we are in the midst of the greatest unravelling of life's rich assemblages since the late Cretaceous period, a sense of urgency and appropriate concern appears generally lacking. In 2021, the 'super year' for biodiversity and host of UN biodiversity conference COP15, negotiations surrounding the UN climate change conference COP26 took centre stage, often overshadowing debates and negotiations concerning the fate our terrestrial kin.

Typically, species which generate headlines are those deemed essential for the continuity of human civilisation (i.e. pollinators), or 'charismatic megafauna' like whales and big cats who have received considerable attention from environmental organisations like the WWF and the SSCS. A culturally pervasive love of butterflies has seen Lepidoptera – the class of insects that includes butterflies and moths – remain the most studied in the world. Spiders, on the other hand – long feared and cross-culturally detested by humans – often remain neglected, a problem exacerbated by negative and/or inaccurate media portrayals of these profoundly misunderstood creatures (Mammola et al. 2020). Aristotle's early denigration of the wasp in relation to bees, to which he assigned the characteristic of divinity (Lehoux 2019), has likely contributed to the longstanding and widespread cultural hatred and fear of these complex social insects (Summer 2022). Similarly, though feral pigeons have shared urban spaces with humans for thousands of years, they too have become the objects of widespread human hatred and disgust since being labelled by officials in the early 1960s as menacing pests defecating on unsuspecting city dwellers. Now, many cities criminalise feeding them in order to control their numbers, and they're often poisoned, shot, trapped, or otherwise destroyed (Jerolmack 2008). However, hopeful exceptions to – and interstices amidst – these hegemonic norms always abound, as when the nineteenth-century French naturalist and entomologist Jean-Henri Fabre mused thus:

> His wing, every feather of which I count, speaks to me of flight among the beautiful clouds; it carries me to the beechwoods whose smooth trunks rise from a moss carpet spotted with white toadstools like eggs left by a wandering hen; it takes me to the snowy peaks, where the bird leaves the starry print of its red feet. My friend the pigeon is magnificent. (Doorly 1942, p. 46)

Fabre was especially in awe of insects, with a palpable sense of wonder pervading many of his detailed descriptions and observations of countless species throughout his life. Unfortunately, invertebrates remain largely out of sight and mind in the realm of public and mainstream concern, for many of the reasons discussed above. A similar plight has plagued snails and slugs, long occupying the top of the Royal Horticultural Society's (RHS) 'garden pests' chart. All manner of lethal and non-lethal methods have been devised by gardening enthusiasts in order to eradicate these mollusc visitors, only some species of which have been found to be detrimental to garden plants. Sarah Ford's *Fifty Ways to Kill a Slug* (2003) is but a testament to this longstanding, exterminative crusade. However, as Andrew Salisbury, principal entomologist at the RHS observes (2022), snails and slugs play vital ecological roles such as helping to recycle dead and decaying plant matter, as well as serving as food for a whole host of other species. He urges a shift in perspective and in how we speak about them, especially moving away from the harmful label of 'pest'. The campaign to eradicate slugs and snails is all the more troubling in the wake of the mounting loss of life that is the defining feature of our present.

Commodified Selves

Many including John Berger (2009) point to the lethal confluence of Anthropocentrism and industrial capitalism as a decisive turning point in modern interspecies relations. Historical archaeologist Daniel Sayers encapsulates this frayed and complex relationship in his aptly titled essay (2014) 'The most wretched of beings in the cage of capitalism'. In the hierarchy of commodified beings generated and perpetuated by capitalism, other animals reside at the very bottom. Now in the Capitalocene, 'they [other animals] are treated as raw material. Animals required for food are processed like manufacture commodities' through the same process that has reduced humans to isolated productive and consumptive units (p. 23) essential for sustaining capitalism's hierarchy of profitability (Marcuse 1969, p. 62). Humans and especially other species find themselves 'entangled' in a reality that continues to wound through a protracted violence (Gumbs 2021). In this vein, Berger follows others in the varied tradition of critical theory in drawing attention to the many socio-ecological deficiencies of capitalism. Perhaps one of capitalism's most insidious effects is qualitative in nature: as discussed in the previous chapter, it turns all, even life itself, into a commodity with exchange value as the sole measure of worth. The late social ecologist Murray Bookchin (1986) perhaps framed it best when he referred to capitalism as a 'cancer':

> My use of the word 'cancer' is deliberate and literal, not merely metaphorical. Capitalism, I would argue, is the cancer of society – not simply a social cancer, a concept that implies it is some form of human consociation. It is not a social phenomenon but rather an economic one; indeed, it is the substitution of economy for society, the ascendency of the buyer-seller relationship, mediated by things called 'commodities', over the richly articulated social ties that past civilizations at their best elaborated and developed for thousands of years in networks of mutual aid, reciprocity, complementarity, and other support systems which made social life meaningful and humanizing. (p. 30)

Ursula Le Guin (2017) uses a similar metaphor when portraying capitalism as a system predicated on growth 'in the wrong sense' … 'not only endless but uncontrolled' (p. 113). As Karl Marx (Marx & Engels 2019) famously theorised, the market economy that is the lifeblood of capitalism works to alienate, exploit and oppress not only the proletariat but also the natural world and other species. What is the lifecycle of a newborn female calf (unprofitable males are often immediately killed at birth) but a succession of inconvenient stages that must be quickly surpassed so that she is ready to be forcibly impregnated in order to produce milk for the agricultural-industrial complex? What good (economically), as Aldo Leopold (1970) poignantly enquired, is an undrained marsh (p. 100)? Such a logic transforms living matter itself into a mere 'apparatus of production' or 'standing reserve' (Heidegger 1977), fodder for hyper-capitalist accumulation (Braidotti 2009, p. 529; Marcuse 2013; Schmidt 2013). 'I-thou' interpersonal relations wherein respect for the irreducible 'other' might flourish are increasingly supplanted by 'I-it' relations (Buber 2000). Within such a system, wherein our social spaces 'saturated with quantitative accumulations of commodities, rather than qualitative modes of relations' (Braidotti 2017, p. 157), any notion of intrinsic value is extinguished in service of an endless series of instrumental transactions with others who are not fellows but, rather, isolated atoms in constant competition with oneself. With profit maximisation as the overriding motive and raison d'etre of this globalised socioeconomic system, other considerations become marginalised. Within late capitalism, therefore, humans appear thoroughly alienated from one another in the labour process as well as from other species, from the land and from the dynamic ecological interactions that sustain life. What's more, and somewhat ironically, within late capitalism traditional human-nonhuman hierarchies are to an extent dismantled through the commodification, exploitation and subordination of all terrestrials – including humans – to the imperatives of the market (Braidotti 2017, p. 165). Hence an array of contemporary ecotopian forms underscore the pressing need of moving towards post-capitalist socioeconomic and political arrangements alongside dismantling anthropocentrism.

Zoocentrism

Though many other species have long been othered and excluded, there has still often persisted an affinity on the basis of biological and other similarities uniting creatures of the same taxonomic category: the kingdom of animalia (Anderson 2009). Members of other kingdoms – plants, fungi, protists and prokaryotes – have been regarded as most distant, categorically denied possession of key biological traits such as learning, memory, cognition and sentience. Mainstream Western science has been thoroughly resistant against granting the latter to plants, still often regarding them as passive organisms (Gagliano 2015; Baluska & Mancuso 2018). Though, emergent concepts such as biologically embodied cognition (BEC) – which emphasises the role of the living body rather than the Cartesian mind in grounding sensorimotor and cognitive processes (Ziemke 2016) – are beginning to challenge this. For plant scientists Frantisek Baluska and Stefano Mancuso (2018), plants are 'cognitive and intelligent organisms which coevolve with humans since the first flowering plants recognised primates as potential frugivores' (p. 51). Plants exhibit their own forms of agency and intentionality – for instance, in phototrophic movement, when they angle and bend towards a light source. Moreover, recent research has found that (at least some) plants emit 'informative' airborne ultrasonic sounds when under distress (Khait et al. 2023). More than this, some plants are known to use alkaloids like caffeine and other less well-understood chemicals in order to 'manipulate the cognition and behaviour of [their] pollinators' (Baluska & Manusco 2018, p. 53). Acacia trees have been found to use their addictive nectar in order to attract aggressive ant species to serve as bodyguards against grazing herbivores as well as for seed dispersal (Grasso et al. 2015). Agency further abounds in mycorrhizal networks which, much like our own neural networks, work to facilitate inter-tree communication through kin recognition, resource procurement and defence (Simard 2018). Even more breathtakingly, some tomato plants, when under attack by mottled willow moth caterpillars, secrete chemicals that turn the caterpillars into cannibals so that they devour one another instead of the threatened plant (Orrock et al. 2017). In light of such sophisticated examples of manipulation, Baluska and Manusco suggest that if there's any case to be made for supremacy vis-à-vis intelligence, it is plants who would be prime contenders (p. 53). Yet, a problematic zoocentrism, or emphasis on the moral significance of *animals*, can be detected in, for instance, some mainstream vegan and animal rights discourses. As with the limitations of focusing on sentience, a focus on animality as the site of ethical significance implies the exclusion of plants as entities worthy of respect and care. The late science fiction and utopian writer Ursula Le Guin, in a brilliantly satirical essay entitled 'A Modest Proposal: Vegempathy', observes:

> Consider, for a moment, what plants undergo at our hands. We breed them with ruthless selectivity, harass, torment, and poison them, crowd them into vast monocultures, caring for their wellbeing only as it affects our desires ... Why, then, if it is immoral to subject an oyster to the degradation of becoming food, is it blameless, even virtuous, to do the same thing to a carrot or a piece of tofu? 'Because the carrot doesn't suffer', says the vegan. 'Soybeans have no nervous system. They don't feel pain' ... that is exactly what many people said about animals for millennia, and what many still say about fish. (Le Guin 2017, pp. 128, 129)

Drawing rigid boundaries around any particular category of beings who matter by default risks designating those who reside beyond as 'less than' or inconsequential. When the spotlight is placed on animals, other beings (i.e. plants and fungi) are either implicitly and/or explicitly consigned to an ethical no man's land. So, the decision to abstain from animal products because of an exclusive ethical concern over the wellbeing of animals – though laudable and beneficial for many long-exploited other-than-human animal kin – risks overshadowing the ecological devastation and systematic destruction of plant life resulting from the production of 'vegan' products such as coconut oil or vegan margarines made with palm oil (which also harm other terrestrials). Moreover, as the excerpt above suggests, arguing that plants don't matter, or matter less than animals – i.e. because they're not conscious or are unfeeling – reproduces the same dualistic and exclusionary logic that has been used to reduce other animals to the status of mindless automatons for centuries. It constitutes yet another example of a conditional ethic that includes entities on the basis of their possession of certain (arbitrary) characteristics, with humans once again as the ultimate judges and arbiters. As I'll discuss in subsequent chapters, it's not the use or consumption of others that's inherently ethically problematic; all organisms use and consume others to varying degrees in the business of life itself, which is dependent upon continual cycles of birth, growth, death and decay. Everyone is at one point or another 'food' for someone else; obviously, we cannot cease to consume others or, as Le Guin (2017) sardonically suggests, subsist on oxygen alone (p. 130). What's ethically unacceptable is the systematic exploitation, displacement and relegation of other species to an inferior status within the system of human supremacy. Abstaining from animal products as an act of political defiance against – and non-compliance with – such systems of oppression is indeed laudable. This is why alternatives to dairy milk, for instance, are worth pursuing – i.e. even more water-intensive forms such as almond milk are often substantially less ecologically destructive (Poore & Nemecek 2018), and ethically considerably less problematic, than dairy milk. However, we need to move beyond the impossible ethics and politics of purity (Shotwell 2016); we are always inescapably implicated in the lives of others, especially amidst the nature-culture entanglements (Braidotti 2020) that mark the Capitalocene. Likewise, all actions have effects that ripple through time and space, and no one – not even the most conscientious vegan – lives without impacting others. Living with others is necessarily messy and complex, involving continuous negotiation, cooperation and conflict. The idea, rather, ought to be to live in ways that minimise systematic harm and devastation, placing an ethos of care, compassion and respect at the centre of how we relate to others.

A Brief Note on Anthropomorphism

Anthropomorphism is the attribution of human-like emotions, motivations and other characteristics to other species (Epley et al. 2007). As Val Plumwood (2007) observes in an imaginatively entitled chapter 'Journey to the Heart of Stone', this tendency can be problematic by enforcing 'segregated and polarised vocabularies that rob the non-human world of agency and the possibility of speech' (p. 20). The dynamic is akin to the silencing of otherness that results from the Western/Eurocentric gaze (Spivak 2005). Similarly, 'human-centered reductionism' (Plumwood 2007, p. 19) in the form of anthropomorphism can involve the arrogant assumption that other species experience the world exactly or very akin to the way we do, thereby overriding their unique and diverse ways of inhabiting the world. It can also be dangerous by ascribing pleasant mental and/or physiological states to other species when, in fact, they might be in distress – for instance, as in the common assumption that the upwards curvature of a dolphin's mouth indicates that it is laughing or content. Moreover, one might erroneously attribute the same avaricious and exploitative motivations and behaviours exhibited by a capitalist human to other species (Despret 2022). In reality we live amidst a multiplicity of worlds, with all manner of beings who live, communicate and interact in ways that are sometimes similar to, and other times quite different from, our own. For instance, as Vinciane Despret observes in *Living as a Bird* (2022): 'anyone who has watched their dog ... rolling in animal droppings will immediately understand that, as far as smells are concerned, we inhabit completely different universes' (p. 23). Whereas, our task should be to multiply worlds rather than, following colonial modalities of inhabitation, reduce them to our own (Despret 2022, p. 29). We might proceed with humility, caution and curiosity, and utilising tools at our disposal (i.e. careful observation, devices to monitor and track other species) which render more visible life-worlds that otherwise would have remained wholly or partially hidden, as the telescope and microscope have revealed minuscule and distant realms previously unimagined (Latour 2004; Despret 2022). In effect, tools that help us *see* better and expand our limited frames. Though, through such explorations it's important to take care that 'when new light is thrown on a situation, it does not then end up obliterating everything under the harsh spotlight of the explanation' (Despret 2022, p. 40). The latter can be tempered by, first and foremost, always proceeding from a standpoint of humility. Then, taking careful consideration of the inferences and connections that we make, of the similarities and differences that we seemingly uncover or surmise, and through the recognition that others always exceed the bonds of our full understanding.

However, one can also go too far in an absolute refusal of anthropomorphism, to the point where one risks rehashing anthropocentric tropes by denying the possibilities of cross-species continuities in knowing, experiencing and being in the world. For, assuming that other species don't or can't experience joy, fear, pain, etc. in similar ways to us is also an exercise in human hubris; indeed, this was the foundation of age-old anthropocentric claims that other species couldn't possibly have similar mental, communicative, social and purposive capabilities to humans. In 'Nature in the active voice' (2014), Plumwood levels a poignant critique of humans' supposed monopoly on speech, the anthropocentric equivalence of communication with verbal language, and therefore the notion that we are the only 'active' speakers. Yet now, centuries later, abundant evidence points overwhelmingly to how – whilst obviously not reducible to one another – there are many similarities in how humans and many other terrestrials feel and experience the world. This is hardly surprising given our shared co-evolutionary histories and common ancestry. Primatologist Frans de Waal makes a useful distinction between 'animal-centred' and 'anthropocentric' anthropomorphism. For de Waal, a fervent denial of human-like thought processes in at least some species can be just as dangerous as anthropomorphising inappropriately (1999). So, one might very reasonably, though cautiously, conjecture – though we can never know for certain – that when a cat actively seeks you out in order to be stroked or sits on your lap, it is probably because they find the interaction a pleasant one and derive contentment from your companionship. When my dog Luna rushes to open the sliding door that opens onto our back garden after I utter the word 'outside', it is because after years of interaction and careful observation she understands the association between that particular utterance and its referent, rather than mechanically reacting to external stimuli. We will never know for certain what our terrestrial counterparts are feeling and thinking, how they experience the world, or what they think of times to come. However, we can and should – with caution and careful attention – attempt to understand and make tentative conjectures regarding what might be going on in their life-worlds. Some degree of this is essential for empathic engagement and ethical relationality. We may not know *what* a crow is saying to us when they vocalise in response to our sudden presence, but we do know that 'she is saying *something* [emphasis added]'; we know that she is 'counting us in' (Blackie 2018, p. 7). To acknowledge as much is an essential first step for forging better ethical relations with our co-terrestrial kin. In her brief and poignant short story 'Mazes' (2012), Ursula Le Guin, imagines how a mouse in a maze enduring endless experiments might experience 'the feeling of being forever watched yet never understood' by a self-absorbed 'alien' species like ourselves (pp. 59–60):

> It remains very hard to ascribe its behaviour to ignorance. After all, it is not blind. It has eyes, recognizable eyes. They are enough like our eyes that it must see somewhat as we do. It has a mouth, four legs, can move bipedally, has grasping hands ... And yet ... the alien has never once attempted to talk with me. It has been with me, watched me, touched me, handled me, for days: but all its motions have been purposeful, not communicative. It is evidently a solitary creature, totally self-absorbed. (p. 59)

As a result of the human alien's persistent failure to attempt to communicate with the mouse, to learn what it needs in order to be properly nourished and content, the mouse concludes, '... now I have to die; but it will not understand the dance I dance in dying' (p. 61). And so, through repeated failures of communication and careful attention, we persist in our mutual misunderstandings. We do a gross disservice to other species both when we assume that their life-worlds are reducible or identical to ours (a violent denial of difference), as well as when we deny the possibility of contiguity (i.e. emotional, experiential), and thereby neglect to try to understand what they might be feeling or experiencing. In other words, it's deeply problematic both to assume total similitude as well as a complete and impassable chasm between worlds; our common origins and co-evolutionary histories have resulted in many similarities across diverse lifeways. What's more, a sense or assumption of total likeness effects a violent erasure of the others' irreducible singularity, whilst assumptions of total discontinuity erect and reinforce perceptions of an impassable gulf between relatives. In contrast, perhaps we might strive to recognise and cherish both – the numerous ways in which our terrestrial counterparts are like as well as unlike us, and work towards establishing a solidarity through diverse experiences and sensibilities.

Expanding Community: Biocentric and Ecocentric Perspectives

Biocentrism, or the 'life-centred' ethic, seeks to overcome some of the issues and limitations noted above by shifting the focus to *life itself* as the seat of moral consideration. Philosopher Kenneth Goodpaster, for instance, remarked that:

> Neither rationality nor the capacity to experience pleasure and pain seem to be necessary ... conditions on moral considerability .... Nothing short of the condition of being alive seems to me to be a plausible and nonarbitrary criterion. (1978, p. 310)

In other words, it's not the capacity for language or even sentience but all *organisms* – from animals and plants to single-celled life forms – who matter (Rolston 2020). American philosopher and nature ethicist Paul Taylor emphasises the 'goal-oriented' nature of all living beings – i.e. all living organisms from plants to viruses and mammals to some degree or another act to promote and/or protect their wellbeing, and exhibit an interest in flourishing. On the one hand, a biocentric approach is more inclusive and, indeed, less arbitrary than selecting rationality or even sentience as a foundation for ethical consideration. However, it rests on yet another contested binary: that between living and non-living, paving the way for non-living entities like rivers and mountains to be overlooked or rendered expendable. Moreover, there is still an ongoing debate within the sciences around exactly how to define 'life', and consequently what does and does not count as a living' entity. Exceptions and surprises continually arise, so that there is no rigid or stable boundary separating the two. Typically, purposiveness, or an autonomous form of agency, and a dynamic kinetic stability are deemed central characteristics of living organisms (Pross 2008). Yet what of an algal bloom, or the most abundant biological entities on earth – viruses? Regarding the latter, Forterre (2010), who defines life as a historical process or 'mode of existence of ribosome encoding organisms (cells) and capsid encoding organisms (viruses) and their ancestors' (p. 151), posits that viruses fit the bill. Moreover, as the social anthropologist Tim Ingold (2006) observes, cultures do not universally discriminate between categories of living and non-living things; many non-Western cultures (see Chapter 6) see 'life' not as emanating from a world that already exists as such, but as 'immanent in the very process of [the] world's continual generation or coming into being' (p. 10). So, dualistic categories such as living/non-living, biotic/abiotic appear to be fluid (decidedly Western) constructs rather than references to fixed or universal ontological realities. Moreover, many of these undoubtedly laudable attempts at including traditionally excluded entities within spheres of ethical consideration still merely extend boundaries. They fail to critically reflect on boundaries as such – on what they include as well as inevitably exclude. If all living beings matter, what of rivers, mountains and all other dynamic assemblages that constitute and are constituted by living entities?

> Ethics is not complete until extended to the land. (Rolston & Regan 1988, p. 188)

Such are some of the queries and ethical quandaries that ecocentrism seeks to address. This ethico-philosophical position, far more than a mere 'analytical abstraction' (Anderson 2010, p. 176), attempts to move beyond the life/nonlife binary by valuing the wider ecological communities that nourish and thereby render possible the lives of individual species and organisms. It is perhaps best encapsulated by the American ecologist Aldo

Leopold, who outlined a 'land ethic' in his seminal work *A Sand County Almanac* (1970) wherein he laments the loss of life and ecosystem integrity in the wake of the spread of industrial capitalism and attendant economic rationality that only values species and ecosystems to the extent that they prove profitable:

> The land ethic simply enlarges the boundaries of the community to include soils, waters, plants, and animals, or collectively: the land ... a land ethic of course cannot prevent the alteration, management, and use of these 'resources', but it does affirm their right to continued existence ... in short, a land ethic changes the role of homo sapiens from conqueror of the land-community to plain member and citizen of it. It implies *respect* [emphasis added] for his fellow-members, and also respect for the community as such. (p. 204)

In other words, ecocentrism entails respect and care for the full spectrum of dynamic entities that make up Terra, from rocks and rivers to those most 'alive' in the traditional sense. It asserts that ecosystems and all of their complex processes and diverse, agentic components matter. As such, it serves as a decidedly more inclusive ethic than zoocentrism's emphasis on animals or even biocentrism's emphasis on *living* entities. Though of course, tensions between ethical duties to individuals and communities always remain and must be negotiated according to each unique context (Rolston & Regan 1988, p. 161). For ecocentrists, the moral community is vast, consisting of rivers, forests, fungal networks, microbes, megafauna, the water cycle, and everything in between. And, crucially, from this perspective the endlessly diverse other-than-human world has *intrinsic value*, independently of their use to humans and whether or not we are around to ascribe value – instrumental, economic or otherwise – to them. This does not mean that humans and other terrestrials never use others instrumentally, only that others ought not to serve as *mere* means to our ends. It is this approach, variations of which have long been championed by indigenous communities the world over, along with critical-posthumanist theoretical and social movements discussed in the following chapters, that I maintain holds radical ecotopian potential. Indeed, as colleagues and I discuss in a recent article (Piccolo et al. 2022), much of mainstream conservation remains decidedly human-centred; The IPBES 2019 global assessment report, for instance, has advanced the concept of 'nature's contributions to people' (NCP) as a means of emphasising that nature's value is 'not just commodities', but to highlight plural and diverse ways of knowing nature and to assess 'all of nature's contributions ... to the quality of life for people' (IPBES, 2021). However, it still does not explicitly avow the inherent value of other species, still framing them as resources who exist largely to further human ends, and focusing predominantly on what nature and other species can provide *for humans*. It asks not what moral obligations we might owe our terrestrial kin, nor how we might help them live the 'good life' too. As such, we argue that an ecocentric turn that shifts the locus of value to all biotic and abiotic components of the biosphere, and the multifarious relations within which they're entangled, would better help stem tides of biological annihilation.

Key developments within the Western scientific canon itself – namely the Copernican revolution which revealed a heliocentric universe, and Darwinian evolutionary theory – have fundamentally challenged many of Anthropocentrism's central tenets. We've begun to see that we are part of an unimaginably vast spectrum of dynamic ways of being in and knowing the world, each unique and suited to a particular spatial-temporal context. What's more, time and again, from cetaceans to corvids, characteristics previously thought to be exclusively 'human' – such as tool use, language, planning for the future and having a sense of self – have been found to be exhibited by our terrestrial counterparts. And this is just the terrain approximating new, tentative knowns and known unknowns; unknown unknowns abound. So, we move the goalpost even further. Nevertheless, even if it could be established that humans have a 'higher-order' mental life and greater intellectual capabilities than other species, as Plumwood (2001) observes, it's unclear how this would mean that other species are inferior. Difference, in other words, need and ought not equate to 'less than'. We likely will never know what it's like for a *bat* to be a bat (Nagel 1974), whether a bat can dream, understand the nature of being, or how it experiences the world, but this doesn't mean that their experiences are any less meaningful than our own. What depths the oxen's gaze conceals will always remain somewhat of a mystery to us – yet this should serve as a source of enduring wonder and fascination. After playing the song 'Les

Cloches de Corneville' to a cockchafer and witnessing no movement from the 'contemplative insect', Fabre remarked: 'Shall we think from this that insects are deaf? ... that would be going much too far. These experiments allow us to say only that an insect's hearing is not like ours' (Doorly 1942, p. 65). If the ability to use echolocation were to be some universal benchmark for ethical consideration and valuation, most species, including ourselves, would miss the mark. Similarly with a dog's incredible olfactory abilities, estimated to be thousands of times more acute than ours. Or consider the mantis shrimp's mind-boggling visual system, with an estimated 16 photoreceptors that allow them to perceive an astonishing array of colours and light forms, including UV and polarised light. Valuing others on the basis of their sameness to us is no foundation for ethics; just as one's skin colour or possession of a uterus shouldn't constitute grounds for green-lighting exploitation, neither should the possession of fur, feathers or scales. Rather, as ethologist Jonathan Balcombe (2016) poignantly denotes, what's beautiful about our other-than-human counterparts, 'and equally worthy of respect, is how they are *not* like us' (p. 233). That is, an ethic and mode of relationality fit for building better beyond the Capitalocene is one that embraces and celebrates everything that makes us *like* as well as *unlike* one another – a unity-in-diversity.

Always Already a 'We'

> Since each of us was several, there was already quite the crowd.
>
> – Deleuze and Guattari (2004, p. 3)

Contrary to what dominant traditions in Western thought have espoused, the atomised, autonomous 'self' is a myth. Prior to the 'Self' and 'other' is the un-atomisable and multifarious 'we' as a 'being-in-common' of singularities (Nancy 2000); the purportedly sovereign 'Self' is *always already* a 'We' (Gergen 2009; Braidotti 2015). Insights from complexity theory (Morin 2007) to indigenous onto-epistemologies (see Chapter 6) reveal that it is relationships with our myriad co-terrestrials, from micro-level interactions to macro-level interchanges amongst global, interactive and dynamic socio-ecological systems – that constitute the foundation of existence (Gergen 2009, p. xxi). Consider the countless microorganisms in our gut that aid digestion and compose part of our skin as the very barrier differentiating the inside (*oikos*) of the 'human subject' from the outside. Hundreds of thousands of fragments (or traces) of ancient viral DNA are integrated with our own (Emerman & Malik 2010). In a brilliant article entitled 'A symbiotic view of life: We have never been individuals', Gilbert et al. (2012) remind us that even our immune system is no mere defence apparatus designed to keep invading pathogens at bay but, crucially, 'mediates the body's participation in a community of 'others' that contribute to its welfare' (Gilbert et al. 2012, p. 333). The thousands of species of microbial symbionts that constitute the human 'microbiome' are thought to outnumber human somatic germ cells by a factor of 10, so that the 'human' is in reality a thoroughly hybrid entity, a composite of microbial and human cells (Turnbaugh et al. 2007). To put things into perspective, approximately 50% of our cells are bacterial, whilst more than 95% of our genes are microbial (Prescott & Logan 2017, p. 9). These microbes help us digest our food, protect us against pathogens, and provide a host of other beneficial services (Prescott & Logan 2017).[3] As such, our own bodies are not even fully 'ours' but rather multitudes of tangled agencies 'that gather together temporarily in fairly durable fashion so as to allow [us] to prolong [our] existence for a while' (Latour 2021, p. 96). 'We' endure thanks to the fortuitous, fragile and fleeting alliances that ensure the necessary conditions of our lives. Such insights deal a further, devastating blow to Anthropocentric assumptions of human separateness and supremacy – indeed, they affirm the ultimate impossibility of Anthropocentrism. They can also provide the foundations for the emergence of new ethical sensibilities. Following Latour (2021) and others, I, for one, am endlessly grateful for the myriad agencies both large and small, historical, present and yet to come, who make 'me' possible. The

---

3   Unfortunately, the sixth mass extinction has not spared our microbial symbionts; they too are in decline due to a host of pressures, particularly due to Western diets which typically lack nutrients, contain excess chemicals, are higher in fat and sugar, and the like (Prescott & Logan 2017).

ongoing COVID-19 pandemic has served as yet another powerful and tragic testament to what can happen when other agencies, in concert with our own immune systems, go awry.

Not only have we never been self-contained individuals, we have never been fully 'human'. Each 'individual' and species is an intergenerational achievement constituted by multiplicities of others across vast spatial-temporalities (van Dooren 2014a). These other agencies operate according to their own laws, to the beat of a different drum. The wider biosphere and its myriad biotic and abiotic components are thoroughly embedded in and constituted by these dynamic multi-actor entanglements (Latour 2004, 2018) that form what Timothy Morton (2010) aptly refers to as a 'mesh' with no absolute centre or delineable edge.[4] There is no 'outside', no space where 'we' *are* and others *are not*, and likewise no definitive line that separates organisms from their 'external environment'. Or rather, something more closely approximating an 'outside' can be said to begin 'where the moon revolves' (Latour 2021, p. 15), beyond the critical, habitable zone that supports and enables the existence of terrestrial life. Individuals and organisms are never self-contained entities interacting with others wholly outside themselves; rather, each constitutes a strand 'in a tissue of [interwoven] trails that together make up the texture of a lifeworld', or 'meshwork' (Ingold 2006, p. 13). The mesh (Ingold 2006; Morton 2010), or 'assemblage' (Deleuze & Guattari 2004, p. 25), is always in flux, in a continuous state of becoming. The innumerable, heterogenous lines and elements that form rhizomes – flowers and their pollinators, us and our viral symbionts – always to one degree or another tie back to one another, rendering rigid dualisms and dichotomies impossible (Deleuze & Guattari 2004, p. 10). These tangled trajectories don't just render boundaries between self and other porous, they are marked by dependencies both up and downstream. The right to be 'individual' can only [partially] be claimed by *perfectly autotrophic* living things [i.e. cyanobacteria] *that leave no residue behind*; [the rest] 'need, at every point of their development, the unpredictable aid

---

4   As Despret (2013) eloquently observes, 'the issue is not about seeking independent existences but about inquiring about the multiple ways one given creature depends on other beings. To be an agent requires dependency upon many other beings; being autonomous means being pluri-hetero-nomous' (p. 44).

of other agents' (Latour 2021, pp. 44–46). Even autotrophs, dependent on molecular symbionts, do not exist outside of relationality. However, us heterotrophs in particular, who cannot generate our own conditions of liveability and must consume others in order to survive, are especially steeped in multiple dependencies.

From such a relational perspective, the sky above no longer appears as the 'divine headquarters' that it did for our ancestors, nor a mere distance to be measured quantitatively, but the 'canopy of an enclosure constantly held in place by the multifarious and multimillennial activity of billions of agencies' (Latour 2021, p. 54). So it is with everything else, from our own immune systems to the climate cycle. Hence the more apt concept of 'intra-action' (Barad 2007), which 'queers the familiar sense of causality (where one or more causal agents precede and produce an effect), and more generally unsettles the metaphysics of individualism (the belief that there are individually constituted agents or entities, as well as times and places)' (Barad in Kleinman 2012, p. 77). However, to acknowledge the porosity and mutability of boundaries, and entanglement between bodies, species, between life and nonlife, is not to descend into a radical monism where existence appears as a homogenous mass devoid of differences (Büscher & Fletcher 2020, p. 130). For there is a continuity between a dinosaur and a bird, 'and yet a dinosaur is not a bird' (Morton 2011, p. 27). Beings grow or emerge out of the interlaced meshworks that constitute their foundation, which is why we can say that relationality comes prior to, but does not negate, individuality. Moreover, it can be argued that there are important (albeit negotiable) boundaries – physical, ethical, relational – that distinguish machines from living entities (Cudworth & Hobden 2013, p. 644), the latter perhaps more fundamentally appearing as singular, irreplaceable intergenerational 'achievements' embedded in complex relations with others (van Dooren 2014a). We are and have always been 'neither One nor Other' (Haraway 2016, p. 98). Troubles arise and abound when one attempts to institute and police rigid boundaries, hierarchies and antagonistic dualisms between 'Self' and 'other', or to equate difference/otherness with inferiority. Ethics – the terrain of how we ought to relate to others – urges that one respect the other *as other* – that is, as an irreducible singularity who always to an extent eludes our grasp, yet with whom we always intra-are.

Making Kin

> We now know what was unknown to all the preceding caravan of generations: that men are only fellow-voyagers with other creatures in the odyssey of evolution. This new knowledge should have given us, by this time, a sense of kinship with fellow creatures; a wish to live and let live; a sense of wonder over the magnitude and duration of the biotic enterprise.
>
> – Leopold (1970, p. 109)

A sense of kinship, as I'll denote throughout this book, is what needs to be (re)established for forging better, more ethical relations with other terrestrials in a world beset by widespread socio-ecological upheavals, which in part can be traced to the rigid binaries and hierarchies that have been central to Western thought for millennia. These modes of orienting ourselves have had devastating consequences for a host of historical 'others', especially other-than-human terrestrials. The related and ongoing legacy of Anthropocentrism underscores Derrida's all-too-accurate observation: 'The worst, the cruellest, the most human violence has been unleashed against living beings ... who precisely were not accorded the dignity of being fellows' (2009, p. 108). We see this in the hyper-nationalistic anti-immigrant fervour brewing in many parts of the world as millions mobilise in search of new places of refuge. And millions more will be on the move in the coming decades as extreme weather events, rising seas, crop failures and resulting geopolitical tensions intensify. Our other-than-human counterparts, relegated to the bottom of the great chain of being, have often fared worst of all. As Crist and Kopnina (2014) observe:

> And herein lies the ultimate irony of the anthropocentric worldview: while it wagered on reaping power by elevating civilized humanity as the apex of creation, and turning the world into its oyster, it has so conditioned human existence that the possibility of an alternative way of life – abundant in diverse beings and rife in mutual flourishing – is virtually beyond thinkable. (p. 392)

As with the necropolitical logic of colonialism and extractive imaginaries, which depict distant people and places as wholly 'other' and thus there for the taking, Anthropocentrism and the wider system of human supremacy reduces whole multiplicities of terrestrials to the status of mute objects who largely matter to the extent that they further our ends. Moreover, the system of human supremacy stifles the imagination and blinds us to other alternatives. Western Anthropocentrism stems as far back as five thousand years, with origins in the humanist tradition in Ancient Greece, famously encapsulated by Protagoras' proclamation that 'Man is the measure of all things'. Thus, it can sometimes appear as though such hierarchical orderings of life are natural or inevitable. Yet, Judith Butler (2017) reminds us of the political promise of abjected bodies consigned to the margins and interstices outside of hegemonic identity constructs, which gesture towards possibilities for a 'radical rearticulation' of the symbolic horizon in which bodies come to matter. Thus, it is the prime objective of this book to demonstrate – by exploring some of today's visionary ecotopian manifestations – that alternative ways of perceiving, valuing and relating to our terrestrial kin have always existed, are always possible and are especially of the essence today. As Weitzenfeld and Joy (2014) remind us, 'Anthropocentrism is not the effect of inescapable, ahistorical constraints of human sensibilities, but rather it is a historic development born from specific institutional and philosophical traditions' (p. 5). Anthropocentrism – along with capitalism, patriarchy, racism and other oppressive systems – is anything but inevitable. We cannot recuperate what has been lost or go back to a mythical 'pure' natural state devoid of our impacts on terra; as philosopher Alexis Shotwell (2016) poignantly observes: 'All there is, while things perpetually fall apart, is the possibility of acting from where we are' (p. 4). We can and ought to, as best we can, find ways of radically scaling back the tides of loss afflicting terra and building arrangements that are fairer for all (Latour 2021, p. 76). We might start by, as Derrida urges, extending 'the similar, the fellow, to all forms of life, to all species' (2009, p. 109) and, again following Haraway, recognising relatives 'as fast and as well as "we" can' (2018, p. 104). We need to (re)build relations with those who have come before us, those still existing and those yet to come so as to forge more habitable dwelling spaces conducive to mutual flourishing in a damaged world. This is the process that Haraway refers to as making 'kin' in a non-biogenetic sense (i.e. not on the basis of genetic or biological relatedness) with all terrestrial beings – including rivers, mountains, and soil. At core this is an ethical task, ethics understood in the Levinasian sense as the pre-cognitive, embodied and thoroughly interactive business of how we treat and relate to others (Levinas 1989). As well will see, this is a task that green utopianism or 'ecotopianism' in its myriad forms, with its transgressive critical and imaginative functions, is particularly well equipped to take on.

CHAPTER 3

A Better World is Reality

> What is denounced as 'utopian' is no longer that which has 'no place' and cannot have any place in the historical universe, but rather that which is blocked from coming about by the power of the established societies.
>
> – Marcuse (1969, pp. 3–4)

The Longstanding Utopian Impulse

'Another world is possible', the slogan that captured the incendiary zeitgeist bursting forth at the 2003 World Social Forum, and of burgeoning anti-neoliberal-globalisation movements (Ehrenreich 2003) the world over. Late capitalism masquerades itself as immutable in a material and ideological sense, thus considerably impeding our abilities to imagine and strive towards alternatives. David Hume famously asserted that one cannot derive an 'ought' from an 'is' (Hudson 1969), but one can critically denounce a deficient 'is' and demonstrate that it needn't be so by highlighting instantiations of the 'better' that abound. Enter utopianism, a multidimensional striving that, in its myriad manifestations, begins with a fervent disavowal of the notion that we as individuals and broader social collectives 'can't behave any other way'. Rather, utopias declare that 'things can be otherwise!', followed by the normative ethico-political assertion that things *ought* to be different. Or, as the utopian critical theorist Herbert Marcuse (1969) observes, a fundamental aspect of utopianism is negation of a status quo perceived to be deficient and oppressive. From these critical beginnings, utopianism goes on to offer creative projections of alternative social worlds through a '"vivid imagination" of the norms, institutions and individual relationships of a qualitatively better society than that in which the utopianist lives' (Pepper 2005, p. 4). Thus, utopianism involves concerted critiques of a deficient socio-political order and imaginative projections of better alternatives. Dystopias, on the other hand, are not anti-utopias – anti-utopianism is most starkly exhibited by neoliberal capitalism which declares that there is no alternative to its bleak world of exploitation and deprivation for the many (Klein 2020). Rather, dystopias' projections of nightmarish alternative worlds serve the similarly critical function of awakening us to the potentially disastrous consequences of continuing along our current trajectories (Sargisson 2012). Dystopias also, crucially, feature critiques of, imaginative challenges, and resistance to the dark society in question (Moylan 2021, p. 99). Dystopias serve as 'a prophetic vehicle, the canary in the cage, for writers with an ethical and political concern for warning us of terrible sociopolitical [and ecological] tendencies that could, if continued, turn our contemporary world into the iron cages portrayed in the realm of utopia's underside' (Moylan & Baccolini 2003, p. 2). And, they are as varied in content, style and structure as utopias – nevertheless, the latter, which seem to be in particularly short supply in the present, will serve the focus of this chapter.

Utopianism's ideological roots in the West can be traced to the mythical 'Paradise' in Judeo-Christianity, a myth that particularly informs many early utopias of the fifteenth and early sixteenth centuries (Dutton 2010). The etymology of the word 'utopia' stems from the famous early modern utopian work by Sir Thomas Moore of the same title (1516) and is derived from a pun on the Greek word *eutopia* meaning 'no place' (Firth 2012). The word *Eutopia* itself is a conflation of two Greek words, *eutopos* which means 'good place', and *outopos*, which means 'no place' (Kumar 1987, pp. 23, 24). The character of 'non-existence' is paradoxically combined with *topos,* a location in time and space (Firth 2012) (a good place) that is not fully realisable *under present societal conditions,* as the excerpt above by Marcuse alludes to. Indeed, it is the 'Event' – i.e. climate chaos – along with a radical new subjectivity that effects a tear in the fabric of the present reality and its normal order of things which in turn occasions the creation of new possibilities previously denounced as impossible by the current order (Badiou & Žižek 2009). Utopia lives in relation between the past, present, and the boundless potentiality of its horizon – the Novum, the new or 'not-yet' (Bloch 1986; Davis 2012). As visions of vastly better worlds predicated on social critique of the 'Now', utopias always have 'one foot in reality' (Linkenbach 2009) and one in the terrain of open-ended potentiality.[1] 'Reality' as such is never 'worked up' but always has something 'advancing and breaking out at its edge' (Bloch 1986, p. 197; Suvin 1990) in a continual becoming. Utopia embodies an ultimately unbridgeable though productive disjuncture between what *is* and what *could* or *ought* to be, which is the site of its transformative potential. In other words, social transformation always remains possible precisely because the 'is' never fully approximates or exhausts what 'could be'.

Lyman Tower Sargent (1994) offers a suitably expansive definition of utopianism as a form of 'social dreaming' (p. 3) that manifests in three key modalities: the more well-known form of utopian literature, communitarian movements (or bioregional utopias/autonomous intentional communities) (Sargisson 2000, 2002[2]), and utopian thought or theory (Bahro 1994; p. 4; Barnhill 2011, p. 128). Of course, these key modalities are not categorically distinct, as utopian theory often influences literary and social movement utopias, and many social movement utopias have been directly inspired by literary utopias (i.e. Starhawk's *The Fifth Sacred Thing* as an influential text amongst anarchist circles). All utopias to varying degrees involve an element of critique, imagine alternative worlds to those within which they're situated, and seek generally to maximise wellbeing. Utopias stretch conceptualisations and contours of the 'now' by placing them in new and previously unimagined contexts (Sargisson 2012, p. 13) (see below). For the prominent utopian theorist Ernst Bloch (1986), conceptualises utopianism

---

[1] Bloch (1986) concedes that the nature of 'the possible' is prescribed by matter itself, wherein real possibility is the 'logical expression for material conditionality of a sufficient kind on the one hand, for material openness (unexhaustedness of the womb of matter) on the other' (p. 206). Thus, under more favourable conditions (structural, socioeconomic, etc.), 'the sculptor' can create 'more beautiful bodies' than the physical ones being born under a certain set of conditions (p. 207).

[2] For more insights on the 'lived utopias' of intentional communities, see Lucy Sargisson.

is a perennial human impulse, a 'reaching' towards better worlds seen in everything from dreams to fully articulated utopias (Sargisson 2012, p. 15). Crucially, for Bloch, utopias are predicated on hope and driven by a 'militant optimism' (1995, p. 200) for the realisation of the 'not-yet' (p. 144). The latter alludes to Bloch's (1986, 1995) 'concrete' utopianism as distinct in method and content from 'abstract' utopianism. Abstract utopias are compensatory, featuring escapism from the exigencies of the 'Now' (Pepper 2007, p. 290), and wherein only the dreamer's position is improved while lacking any further engagement in advancing societal transformation. Concrete utopianism, on the other hand, is derived from critical social theory (Petersen 2010, p. 18) and helps us sharpen our critiques of existing society through a critical engagement with the unbounded horizon of possibility surrounding the real (Bloch 1986, p. 223). Thoroughly rooted in a life-affirming and collective praxis (Braidotti 2013; Dinerstein 2017), it involves concrete action for bringing about an anticipated and desired 'Not-Yet' here and now.

For Mannheim ([1939] 2013) there is a crucial distinction to be drawn between 'ideology' and 'utopia', wherein the latter is essential to humanity's continued capacity to 'shape and understand history' (1939, p. 236). Those visions or sets of ideas which appear to transcend present reality yet ultimately integrate themselves into the present, perpetuate and reinforce the status quo – whose dominant ideology proclaims the inevitability of neoliberal capitalism (Newman in Davis & Kinna 2009, p. 208) – remain mere ideologies. Utopias, on the other hand, emerge when social groups envision and put their ideas into practice, thereby shattering the existing social situation (Mannheim 1939, p. 192). As such, for Ricoeur (1986) the liberatory utopian imaginary is the very antithesis of conservative ideological thought. This recalls the Blochean 'concrete utopia' – or the prefigurative political modality which provides 'free spaces' (Sargisson 2002; Barnhill 2011) within the existing actuality wherein one can enact utopian principles (Pepper 2005, p. 4). Of course, in a sense every ideology contains elements of utopianism (Sargent 2006), and similarly, utopias risk losing their radical, critical function when they become co-opted into existing systems and ideologies (Miéville 2015). Though, there is an important distinction between Mannheim's (1939) and Bloch's (1986) conceptualisations of the 'real' – the latter in particular, as previously denoted, conceives of the 'real' as a perpetual process of becoming and thus of utopia as an always-possible future(s) as a reflection of the always-unfolding nature of reality (Levitas 1990, p. 19). Some (Kumar 2003, p. 69) see approaches to utopianism like Bloch's (1986) as too broad, contending that utopias, in dealing with possible rather than actual worlds, are always and only fictions that can be associated with a certain set of generic conventions (Sargisson 2012, p. 15). However, along with others (Levitas 2003; Clark in Davis & Kinna 2009, p. 23) I maintain that to reduce the utopian imaginary merely to the realm of literary fiction is 'to flatten out its richness' (Sargisson 2012, p. 16). Utopianism manifests in a multiplicity of forms attendant to the plural nature of reality as such, and whose destabilising functions – in the form of radical critique and imaginative projections – harbour considerable transformative potential. Much of utopianism's functional value lies in the putting forth of counterfactual experiments designed to critically 'gauge the dispensability of various social factors' and demonstrate that it is indeed possible to create societies without extreme socioeconomic inequalities and rampant environmental degradation.

Utopia and Its Discontents

> In a modern Utopia there will, indeed, be no perfection; in Utopia there must also be friction, conflicts and waste, but the waste will be enormously less than in our own world.
>
> –Wells (1905, p. 176)

Some popular conceptions and anti-utopian academic discourses (Gray 2007; Popper 2014) have variedly framed 'utopia' as a hopelessly idealistic striving for impossibly perfect worlds. In this vein, utopianism is typically dismissed as a mode of thinking or projection of alternate modes of living that are so 'idealistic' as to reside beyond the bounds of conceivable practicability. This anti-utopian sentiment is akin to portrayals of human nature and society as too irremediably violent and self-regarding to ever be conducive to the implementation of better modes of being (Hobbes

2016). In his collection of reflective essays *Some Thoughts on the Common Toad* (1984), George Orwell offers a detailed and critical assessment of Jonathan Swift's seminal literary work *Gulliver's Travels* (1726), a satirical and critical expose of eighteenth-century England. Orwell unearths a distinctly anti-utopian undercurrent in Swift's work which implies that this world cannot be substantially improved and that humans are irremediably 'wicked' (p. 58). He draws on Swift's depiction in Book 4 (a book also focused on interrogating human relations with other species) of an ideal society presided over by the Houyhnhnms, a species of intelligent horses. Despite Swift's scathing critiques of totalitarianism, Houyhnhnm society exhibits a rigid (racialised) caste system wherein 'reason' has become a totalitarian organising principle aimed at purging all contrary passions and interests from their dealings with one another. Orwell thus observes:

> The most essential thing in Swift is his inability to believe that life – ordinary life on the solid earth, and not some rationalized, deodorized version of it – could be made worth living. Of course, no honest person claims that happiness is now a normal condition among adult human beings; but perhaps it *could* be made normal, and it is upon this question that all serious political controversy really turns. (p. 51)

To be labelled as 'utopian' is often to be dismissed as delusional, hopelessly lost in a haze of imaginative speculation and therefore out of touch with how things actually and necessarily 'are' (Buber 1996). This is, after all, what 'capitalist realism' (Fisher 2009) continually (re)asserts: that there is no – and, indeed, there cannot be any – alternative to capitalist modes of production and consumption under the sun. Yet as Orwell (1984) rightly denotes, like all animals, humans are behaviourally and motivationally complex, often juggling self as well as other-regarding interests simultaneously, with anti-social as well as prosocial tendencies amenable to transformation. To suggest that this world only has ruthless oppression, exploitation and widescale destruction in store is a gross simplification and arrogant reduction of reality's pluralistic nature and unbounded potentiality (Orwell 1984, p. 60). Moreover, such a claim at least partially appears to be rest on the common assumption that utopianism aims to yield individuals and societies that are perfect, unchanging, and unerring, which, historically and particularly in utopianisms more contemporary manifestations, is not the case (Sargent 2006).[3] Indeed, Zamyatin's classic dystopia *We* (1993), which inspired George Orwell's *1984*, powerfully underscores the undesireability – and dangers – of a society that has ruthlessly implemented its teleological vision of a perfect 'end state' wherein mathematical precision, calculation and control underpin everything from human-nature relations to love and one's innermost desires. The society depicted in *We*, presided over by OneState, is 'cloudless', straight, and totally unerring (Zamyatin 1993, p. 65). Hence why it is classed as a dystopia. By contrast, especially with more contemporary modalities (see below), utopianism (at least at its best) seeks to bring about social relations and structures that are not perfect, but considerably better than extant forms. Utopias understandably wish to preserve their instantiations of 'the better' overall and safeguard them against radical alterations, but this does not mean that they are resistant to change. Utopianism – drawing on hope, as opposed to faith – asserts that a better (not perfect) world is *possible*, not guaranteed (Solnit 2016). Moreover, whatever the extent of human and societal malleability, and the varying influences of agency versus structure, the fact that insurgent realities like Earth First! and Fridays for Future exist demonstrates that *substantially different* modes of being can be brought about. Utopianism is therefore a decidedly realist political project which draws on the always-unfolding potentiality that characterises reality as such as well as the 'Not-Yet' (Bloch 1986); to deny such potentiality is the truly 'unrealistic' act.

Others, often alluding to historical examples such as the USSR and People's Republic of China, have claimed that utopianism's pursuit of human and societal perfectibility – in effect the attempt to impose a particular conception of truth or 'the good' – bears the seeds of totalitarianism and intolerance of difference (Kolakowski 1983; Popper 2012). Indeed, inklings towards domination can be seen within *some* totalising/blueprinting utopias and their cultural-historical links to Enlightenment projects of

---

3   As Sargent (2006) poignantly observes, 'Thomas More did not pretend that the society in his Utopia was perfect; Edward Bellamy's eutopia Looking Backward is changing, and did change in that he wrote a sequel that had differences. H. G. Wells says that the eutopia in Men Like Gods is rapidly changing. The only utopias that I can think of that can be said to be perfect are some myths of an earthly paradise and some of the depictions of heaven that were popular in the late 19th century' (p. 13).

universal conquest, namely the hegemonic 'utopias' of Western modernity (Latour 2017) and more recently of neoliberal capitalism (Clark in Davis & Kinna 2009, p. 12). Among the most notable critiques of utopianism can be found in Marx and Engels' writings, particularly in the latter part of their *Communist Manifesto* (1848).[4] Engels, for instance, acknowledged the great debt owed by German socialist theory to the visions espoused by classic utopian socialists such as Charles Fourier, Robert Owen, William Morris, and Henri de Saint-Simon (Buber 1996). In some instances, Marx deploys decidedly utopian language when delineating an emergent communism that 'proves its possibility in the teeth of the widespread notion of its "impossibility"' (Buber 1996, p. 87). Nevertheless, Marx and Engels appeared to be especially critical of utopian socialists whose imaginaries – despite containing notable critiques and insights into the nuances of the capitalist exploitation – they saw as groundless abstractions divorced from 'scientific' critiques of political reality and participation in the struggles of the working classes (Buber 1996; Engels 1999; Davis 2012, p. 128). Marx, Engels, and Lenin provided erudite political-economic critiques of the gross deficiencies of capitalism – the first step in utopian thinking – yet fell short of the second in a curious refusal to creatively delineate the contours of what *could* come next (Landauer 1978). Nor did they sufficiently direct their gazes towards the potential building blocks of better worlds – i.e. producer and consumer co-operatives – that existed and continue to do so within the present. Lenin is purported to have remarked, 'what socialism will look like when it takes on its final forms we do not know and cannot say' (quoted in Buber 1996, p. 115). Yet, while we may not know precisely what new societies will look like, we can work towards a general consensus on what we *want* them to look like, and thither direct our efforts (Buber 1996, p. 115).

Utopianism in all of its manifestations has often arisen during times of significant socioeconomic and political upheaval as a response to the concrete needs and experiences of segments of society most threatened by the dislocations wrought by deleterious social, economic, and political

---

4 Even though one can detect palpable utopian-socialist elements in their works and an overall fluctuating and ambiguous relation to utopianism.

conditions. Likewise, the utopian socialism that blossomed throughout the first half of the nineteenth century was a response to the rapid development of industrial capitalism which led to uneven social dislocations, a burgeoning working class divided between a proliferation of occupations, and a steady erosion of traditional pre-capitalist social values (Bookchin 2005). The turbulent social climate of the time created a powerful, though by no means homogenous, impulse amongst utopian socialists like Henri de Saint-Simon and Charles Fourier (Taylor 2013) towards a more harmonious social order. These can hardly be dismissed as mere groundless abstractions and/or idealist fancies; their articulations of alternatives, vital for guiding societies towards better 'Not-Yets', are always born in response to concrete experiences of lack and injustice. And many forms, crucially, assert that a better world is not a given but that it is possible through collective action. Recall our previous discussion of Bloch and the notion that reality as such is never fully worked out or determined; as such, a better world – or, conversely, a worse world – is *always possible*, but *never guaranteed*. What has since persisted as rather intractable 'strategic dilemmas' within utopianism more broadly concern its process (how we get there) as well as its content (what ought it to look like?), a endeavours rendered especially complex when the conversation is expanded beyond the human! Thus, questions remain as to whether new social orders ought to be industrial, anti, or post-industrial, feature private or communal ownership of property, religious or secular, achieved gradually or through revolution (a key point of Marxist contention), or organised statist-authoritatively versus communitarian-democratically (Goodwin & Taylor 1982).

Utopian Conventions of Critique: Play, Estrangement and 'Otherness'

All manifestations of utopianism – especially in literature and media – employ certain conventions and tactics for highlighting the inadequacies of the now and articulating alternatives, such as excess and play. Excess[5] as

---

5   Just as the Earth and others are always in excess of- and therefore place limits upon- our own worlds.

the very *beyond-within* the living present (Derrida 1999a) is often manifested by the individual or character 'playing the fool'. This character often sees what others cannot and exposes the 'truth' of the inadequacy of the now (Sargisson 2012), thereby permitting the flow of 'radical creativity' and allowing for the positing of wildly different alternatives. Through play (with conventions and norms of the society under critique), utopians 'fool around with reality ... like a dog with a rag ... and tweak the nose of convention: transgressing norms, breaking rules and crossing boundaries' (Sargisson 2012, p. 16), demonstrating the ridiculousness of life as it is and most significantly, that things could be otherwise. Another common convention is the presence of 'the visitor', a character who usually comes from the same spatial and/or temporal context as the author who can view the utopian or dystopian world through the critical gaze of the unfamiliar other. The visitor serves the vital function of contrast, permitting dialogue to occur between the author's present and the radically different 'other place'. For instance, in Edward Bellamy's *Looking Backward: 2000– 1887* (2010), protagonist Julian West, who plays the role of the visitor, finds himself in his native Boston 113 years into the future wherein a new socialist utopia has emerged. Upon asking on what basis the utopian state distributes sustenance 'credits' to the workers, his host Doctor Leete responds: 'His title ... is his humanity. The basis of his claim is that he is a man' (p. 60). Incredulous Julian then learns that all citizens receive the same level of basic sustenance regardless of their occupation (an early precursor of contemporary universal basic income schemes). This nascent realisation is made all the more striking by virtue of its stark contrast with the capitalist Boston of Julian's birth (and the society under Bellamy's critique), wherein scores squander in abject poverty while a mere handful revel in reckless extravagance. Thus, in the juxtaposition of the radically 'other' – and markedly better – utopian world alongside the deficient society effected by the presence and contemplation of the visitor, the deficiencies of the old and the redeeming qualities of the new come into light.

The key convention of estrangement (Suvin 1990; Sargisson 2007) is especially important, practically as well as conceptually, in literary and non-literary manifestations of utopianism. Estrangement is not only a utopian method of critique but, crucially, touches on central dynamics such as self/

other, inside/outside, etc. In its relation to other key concepts such as difference and alterity, and interspecies relations more generally, it is also worth further discussion. The term 'estrangement', which contains cognate terms similar to distance and difference, traditionally has evoked notions of loss and nostalgia, as with an 'estranged' lover or family member with whom one has parted ways. Conceptually, however, the notion of 'estrangement' features a complex amalgamation of various modes of distance – normative, ideological, social, etc. (Sargisson 2007, p. 394). In utopias, these modes of difference evoke the stranger and the extraneous by wilfully placing a person or thing 'outside' of the familiar. Estrangement is essentially what permits utopias in their myriad forms to function critically, as radical critique must to a degree be predicated on separation from the object of critique (industrial capitalism, for instance). Through the distance generated between the stranger and the 'boundaries of the known', which can be spatial, emotional, temporal, ideological, political, cultural, or onto-epistemological, utopias create the necessary conceptual and physical space for critical reflection wherein critiques of the 'now' and radical visions of alternate worlds may spring forth. Through the transgressive nature of play, estrangement, satire, and the role of the visitor, utopias both in literary and in sociological form challenge the ways we think by showing established institutions and conventions to be what they truly are: porous, contingent, always subject to change, and in Latourian (2004) terms, dependent on the strength of their alliances in order to retain their powers of influence.

Whilst an essential feature of utopianism, estrangement can also be problematic if excessively antagonistic or if otherness as such is construed as undesirable (Plumwood 2002). Put another way, as discussed in the previous chapter, ethically problematic is the dialectical approach to difference which construes difference or otherness hierarchically as an object to be governed or exploited (Braidotti 2013, p. 68). For instance, Sargisson's (2007) ground-breaking work with green intentional communities reveals that estrangement in community form can also often be physically (distance from basic amenities such as schools and hospitals) and emotionally taxing (due to the intensity of close-knit interpersonal relations within small communities). Excessive physical-psychological estrangement of utopian spaces from the 'outside world' can engender a form of 'collective alienation' within the 'Otherness' of utopian spaces themselves (Sargisson 2007). Herein, in response to the 'hostile gaze' of the world under critique, the utopian mode or community in question becomes increasingly introverted, engendering a regressive parochialism wherein the initial vision can become further radicalised or distorted. Such mechanisms in turn risk heightening the utopia's sense of Otherness as well as ignorance and/or fear of the 'outside' (Sargisson 2007, p. 403). While this may strengthen in-group bonds, it may also heighten hostility towards 'outsiders' deemed threatening to the in-group's way of life. In other words, such a dynamic may ossify boundaries to the extent that the 'other' is regarded with hostility and utterly excluded (Derrida 2005b). This can also undermine the utopias' ability to effect wider societal transformation beyond its borders. This is why some (Morton 2010) detect potentially fascist tendencies in the concept of a 'community' and its allusions to an 'inside' as fundamentally distinct from an 'outside' (p. 278). The finite nature of our ethical communities is such that there is always an excluded 'other' (Derrida 1995b, c), and it is important to remain aware of the ethical and political ramifications of this. While necessarily radically distinct from the 'other', the 'outside', and/or 'the now', and though estrangement can be a powerfully liberating force, utopian spaces must also take care to avoid regressing into a detached hostility towards the 'now' that can impede their wider critical functions. Estrangement in excess can be counter-productive; boundaries, while indispensable, must be selectively permeable and continuously interrogated, more mesh or membrane rather than solid concrete (Sargisson 2007; Morton 2010).

Key Utopian Concepts: Hope, 'Reality' and Futurity

Hope is a central concept in the utopian tradition, with a complex relation to the 'Now' and 'futurity'. Hope can be defined as a future-oriented emotion or belief stemming from *concrete* dissatisfaction with life as it is at present (Lazarus 1999) and containing exhortations that things *ought* to change (Baumgartner et al. 2008). For philosopher Jacques Derrida

(1999a, 2005a), hope contains the 'invincible elán or affirmation of an unpredictable future-to-come' – as the very site of justice (1999a, p. 253) – whose paradoxical presence and non-presence is experienced as the singular urgency of a 'here and now'. This hope-as-emancipatory-desire serves an invaluable political function in the present by sustaining 'militant and interminable political critique' (Derrida 2005c, p. 86). The future – like the 'other' – can never be fully anticipated or known by finite and historically situated subjects. Thus, for Derrida and in relation to futurity, our responsibility[6] of absolute openness to the 'other' precludes any possibility of ascribing definite content to the latter in the form of, for instance, an ecotopia-to-come. His notion of 'messianic without messianism'[7] refers to the very structure of the promise of a future which is always to-come, without determinate content, and whose radical otherness demands that we maintain a complete openness to it. Such a reluctance to attribute content to or strive for a particular conception of the 'good' is said to be necessary for preventing a descent into totalitarianism. This radical openness recalls some utopian modalities which resist closure around a broad set of desired goals and principles (see below), in a related resistance against ascribing definite content to the to-come. However, I would emphasise that ethics – the terrain of how we ought to treat/relate to others – demands *some* form of intervention in situations of intolerable injustice (i.e. biological annihilation). We inevitably must (tentatively) strive towards some idea of the better (i.e. a world without extreme

---

6   'Responsibility' for Derrida (Diprose 2007) is inherited; as a historically situated and finite being, the subject's inheritance of society's moral customs and norms are necessarily partial and unique. Thus, in contrast to the Nietzschean 'sovereign subject', for Derrida subjectivity *is* responsiveness to and responsibility for the other; however, though no one is wholly free or sovereign, 'a certain space of freedom is opened' by responsibility that would allow something other to arrive, that would allow 'something of the future-to-come' (Diprose 2007, p. 444).

7   For Derrida (2005a), the structure of the 'future-to-come' is absolute futurity and irreducible singularity, yet it also- and somewhat paradoxically- has the 'hardness, closeness, and urgency' of the real' (2005, p. 131); hence his understanding of messianicity as an eminently real and concrete event (p. 131). The experience of the 'to-come' is that of the 'singular *urgency* of a *here and now*' that is in a sense both with and without presence (2005b, 2005c).

socioeconomic inequality or the mass exploitation of nonhumans) if we are to alleviate injustice. Likewise for Bloch, justice cannot take the form of an empty messianic promise but must be rooted in praxis, brought forth by collective strivings for justice.

Deliberations surrounding the immanent (here and now) versus transcendent (the beyond) nature of utopianism are longstanding, though it is the former understanding that has proliferated in recent decades and which is emphasised throughout this book. Many, including those sympathetic to the utopian tradition, have contended that utopianism on the whole engages in the systematic blueprinting of perfect societies, depicted as static and resistant to change, and therefore bearing the seeds of totalitarianism (Kumar 1987; Davis 2012). Many utopias (Campanella 1602; More 1912; Bacon 1915) have indeed been characterised by certain degrees of stasis, disembodiment from concrete transformation of their 'present', and evincing a seeming disavowal of the unbounded potentiality of the 'to-come'. Owing to utopianism's Christian roots, 'transcendent utopias' (Davis 2012, 2021 seek to evade the difficult struggles of earthly life and find eternal peace in 'some kind of Heaven or Nirvana'. These often feature a dichotomous opposition to their present reality, depicting time and historical movement as linear successions of stages proceeding from an archaic past and towards a future wherein the utopian ideal is seen as occupying a 'fixed space outside time and history' (Davis 2012, p. 132). Such spatio-temporally 'transcendent' (Davis 2012) or 'future-oriented' utopias marked the dawn of the modern phase of utopian thought, originating in the seventeenth century alongside new theories of social evolution which coalesced into modern Enlightenment notions of linear historical progress[8] (Kumar 2003, p. 67; Garforth 2009, p. 12; Linkenbach 2009, p. 3; Petersen 2010, p. 15). Traditional notions of society and the human condition as irremediably war-like (Hobbes 2016) were considerably undermined by the Enlightenment underpinnings of modern utopianism. A new faith in

---

8   This distinctly Western phenomenon stems even further back, towards the end of the seventeenth century, with the introduction of the modern historical time periods 'ancient, medieval, modern'. Rather than existing in dialectical relation to one another, the past was newly conceived as a 'closed case' wholly separate from the 'present' and the 'future' (Davis 2012).

reason and applied science engendered a more fluid, 'evolutionary image of society-in-time', along with 'the perception of society as malleable' and of social change towards the better as possible (Levitas 1979, p. 26; Goodwin & Taylor 1982, p. 143). This marked a crucial ontological shift which paved the way for nascent conceptualisations of utopianism as a catalyst for societal transformation, wherein the 'human' and the 'social' were no longer seen as fixed or irremediably corrupt. A sense of optimistic faith in social, material and scientific progress followed suit, along with emphases on humanity's growing abilities to manipulate the natural world due to burgeoning scientific discoveries in such fields as ecology in the twentieth century. The latter is particularly palpable in a number of Wells' (1905, 1914, 1923, 1933) works published around this period (Alt in Canavan & Robinson 2014, p. 25; Alberro 2022).

The dynamics of spatial-temporality in utopianism vary (Levitas 2003) between utopias of temporal process and utopias of spatial form – the latter often depicted as ideal locations dislocated in space, such as in More's *Utopia*.[9] For Marxist geographer David Harvey and his work, *Spaces of Hope* (2000), the social production of space is of central importance in utopian deliberations, not least because it is a fundamental feature of capitalism's instantiation – through its steady colonisation of the commons, its paradoxical production of limited spaces and the need for their destruction in service of its continuous geographical aggrandisement for capital accumulation. Similarly, capitalism requires a 'spatial fix' (Levitas 2003) in its externalisation of the social and ecological costs of production, as well as in its generation of elite utopias in the form of gated communities and luxury doomsday bunkers (Stamp 2019) as refuges for the ultra-wealthy amid burgeoning dystopian surroundings. Hence, as explored below, the emergence of critical and grounded utopias intent on creating spaces of radical alterity within and against the spatial dislocations wrought by global capitalism. These spaces serve the vital (utopian) function of disrupting the status quo and/or influencing it by way of embodying preferable alternatives. However, it should be added that spatial utopias also risk losing their

---

9   This is largely because More's *Utopia* (1516) is an early-modern text preceding Enlightenment ideals of linear progress.

radical and subversive potential through compromise with and co-option by the dominant society (Levitas 2003, p. 140). Moreover, as with previous categorisations (i.e. abstract vs. concrete) utopias never feature merely spatial dimensions with no traces of the processual, and vice versa. Hence the allure of a more 'dialectical utopianism' (Levitas 2003) that neither exclusively prioritises spatial form nor temporal process in the Marxist-Hegelian tradition, but rather foregrounds ongoing processes of struggle in particular spatial localities and their relation to social structures, while also gesturing towards better alternatives. Binary oppositions aren't useful, as reality is always far messier, defying attempts at clean categorisation, as do utopian manifestations in their myriad forms. All contain elements of process and spatiality; all contain complications and paradoxes. What is increasingly apparent amidst the exigencies of the Capitalocene, however, is that the here and now requires urgent and concerted attention, as a space for simultaneous resistance and the embodiment of alternatives.

Time: Homogenous, Dialectical, and Disjointed

> ... it is necessary to reject the notion of time that began in Europe during the eighteenth century and is closely linked with the positivism and linear accountability of modern capitalism: the notion that a single time, which is unilinear, regular, abstract and irreversible, carries everything. All other cultures have proposed a coexistence of various times surrounded in some way by the timeless ... Every one of the crossing energies operating in a forest has its own timescale. From the ant to the oak tree. From the process of photosynthesis to the process of fermentation.
>
> – Berger (2007, p. 139)

For Latour (2004) the 'modernist' conception of time as a linear progression from an archaic past to the 'utopia of a radiant future' (p. 195) is predicated on untenable distinctions such as between Subject/object, Society/nature, and Facts/values (p. 188); without the help of Scientific rationality with a capital 'S', there was no hope for salvation from the dual threats posed by the brute necessity of an external 'Nature' and the barbarism of unsubstantiated beliefs (p. 188). Yet, as we'll explore further in Chapter 6, linear notions of historical time are far from universal, but are decidedly Western (and human) phenomena. Enduring debates on whether or not there is such a thing as continuous time in any objective sense have problematised notions of 'linear progress' and engendered a resurgence of caesurist thinking which 'eliminated the continuity between the present and utopia' (Levitas 1982, p. 57). An illuminating early example of this can be traced to millenarian eschatological movements which, often arising during periods of considerable socio-political precarity and uncertainty, expect the "'complete destruction of existing social, political, and economic order, which will then be superseded by a new and perfect society'" (Barkun 1986, p. 18; McNeish 2017, p. 1039). In more explicitly religious incarnations, the bringing forth of the new society after a radical break with the old often involved divine intervention (Levitas 1982). In more secular versions this would be achieved by the masses themselves through collective political action, for instance. A cultural-historical corollary can be found in the ground-breaking theory in evolutionary biology of 'punctuated equilibrium', proposed by Stephen Jay Gould (Eldredge & Gould 1997). The theory of 'punctuated equilibrium' challenged a core assumption in Darwinian evolutionary theory by positing that species (and natural systems) evolve not only gradually but, also, through alternating periods of relative equilibrium 'punctuated' by periods of 'revolution' – i.e. significant upheavals in the form of mass extinction events (Bak & Sneppen 1993). Such cataclysmic events tend to coincide with 'adaptive radiation', a proliferation of new or previously sparse species, like when mammals mushroomed in the time preceding and following the annihilation of dinosaurs at the end of the Cretaceous period. Contemporary (green) utopias, which operate amongst such 'pessimistic, apocalyptic or pragmatic tropes' (Garforth 2018, p. 23), already exist within the 'end times'; thus, a crucial question that I will return to in the final chapter is, what happens to hope and conceptualisations of the 'to-come' in the dark (Pohl 2019)?

As discussed, utopianism harbours many complexities and inconsistencies surrounding the presence and non-presence of the 'Not-Yet' in relation to 'reality'. Bloch (1986) occasionally refers to the tendencies present in the terrain of the utopian 'Not-Yet' as the 'accumulating pressure of different modalities of potentiality and possibility on what has become' (Anderson

2009, p. 700). Bloch (1986) conceptualises the 'Not-Yet' as the *presence* of a fragment of something better residing on 'the front of the world process … of animated, utopianly open matter' (Bloch 1986, p. 200; Anderson 2006). This is why the 'Not-Yet' as well as the 'Now' is always irreducibly 'other' to an extent, because one can never exhaust their infinite potentiality (Bloch 1986) nor succeed in 'mapping them out' in their entirety. This disjuncture between being and non-being, between things as they presently have materialised and as they *might yet* be, is the source of utopian refusals of the 'Now'. Latour (2004) rejects exclusively future-oriented utopian projects in their purported disregard for the complexity of the present, emphasising the need to focus on co-constructing our common world in the 'Now' alongside others with whom we share Terra's habitable zone (Morin 2015). Bloch's (1986) concrete utopianism similarly functions as an immanent disruption of the 'Now' while attempting to create radical spaces of the 'better' within it. Echoing Latour's actualist ontology, Bloch (1986) refers to the utopian horizon 'continuously included in the reckoning', wherein 'the real appears as *what it is in concreto*: as the path-network of dialectical processes which occur in an [always] unfinished world' (p. 223). The 'animated' and 'utopianly open matter' is developed through the concrete embodiment of hope through praxis for the instantiation of better alternatives.

Utopianism conceived thus as the concrete unfolding of 'open matter' marks a shift from traditional notions of utopia as either abstract configurations located in a distant future or already latent in full within the 'Now'. Moreover, in this always-unfinished world interrupted by the excesses of utopian potentiality, cracks always appear sooner or later, and it is in these spaces (as discussed below) where new alliances and assemblages can take root, gathering the necessary strength to displace the old. In his seminal work *Negative Dialectics* (1973), Frankfurt School critical theorist Theodor Adorno muses on our present, damaged life and avows utopia as a 'still to come' wherein the very ideology of domination will no longer be, wherein humans will be reconciled with nature and their wider terrestrial kin; that is, a utopia 'above identity and above contradiction … a *togetherness of diversity*' (emphasis added) (p. 150). As Adorno further observes, although identification has the potential of generating awareness, it also has a tendency of positing the self in the other and thereby overriding the latter's singularity, which in turn can pave the way for domination. As discussed in Chapter 2, to assume full access to and knowledge of the 'other' – nature, other species, the future – constitutes a form of domination. Rather, the aim ought to be to strive for a togetherness of diverse, singular beings-in-common. Adorno offered erudite analyses of the ways in which domination in late capitalist society has permeated every fibre and dimension of contemporary social life (Martin in Biro 2011, p. 123); however, following Latour and Bloch, he also saw fragments of potentiality along the margins and in the interstices wherein resistance to capital's hold has been flourishing all along.

Pluralistic Strivings: Critical Heterotopias of the 'Here and Now'

> What kind of life? We are still confronted with the demand to state the 'concrete alternative'. The demand is meaningless if it asks for a blueprint of the specific institutions and relationships which would be those of the new society: they cannot be determined a priori; they will develop, in trial and error, as the new society develops. (Marcuse 1969, p. 86)

Tom Moylan's (1986, 2021) notion of 'critical utopianism', which emerged during the 'historic block' of political opposition of the 1960s and 1970s (Breines 1989), marks another important step within the utopian canon towards praxis and the essential 'Nowness' of critique and imagination. This 'critical' strand of utopianism is fuelled by a fervent opposition to the exploitation and domination of people and the natural world increasingly characteristic of Western capitalist societies (Breines 1989, p. 25). This opposition is coupled with a utopian desire for mutual aid, ecological resilience, liberation, and peaceful living (Moylan 1986, p. 11) – in effect, a longing for community (Breines 1989, p. 26). The 'critical' in Moylan's (1986) conception of the 'critical utopia' is twofold: foremost (1) in the sense of *critique* through debunking and deconstructing the utopian genre itself, namely its colonisation by capitalism towards the 1970s and infusion with technocratic dreams of material abundance through boundless economic expansion; and (2) in the sense of the *critical mass* required to effect fundamental societal transformations (p. 10; Sargisson 2012). If the better is the enemy of the good, then perfection serves as 'a stick with which to beat the possible' (Solnit 2016, p. 77). Thus, critical utopias do not offer rigid blueprints of a fixed or perfect ideal world but rather head *towards 'the ideal'* by offering 'a rich blending of creative fantasy, critical thinking, and *oppositional activism*' predicated on a demand for radically new relations between social and natural systems (Moylan 1986, p. 27; Sargisson 2012, p. 11). Ever critically self-reflective, critical utopias continually remind us of the limitations of the utopian impulse, particularly the dangers of pursuing rigid and transcendental visions of 'the better' (Moylan 1986, 2021). Nevertheless, these are desperate times; time is of the essence. In his impassioned work *Becoming Utopian* (2021), Moylan explores the trajectory of radical utopian subjects' break with the established order and embrace of the boundless potentiality of the to-come. Reflecting on left tensions around desirable modes of political organisation, he gestures towards the need for an intersectional, democratic unified leadership structure that can help sustain a political movement for counteracting the 'global/neoliberal superpower order while reaching for a utopian horizon' (p. 13). With a focus more on process than a fixed ideal, the critical utopia explores and articulates vibrant post-capitalist futures that are simultaneously amorphous, open-ended, and far more socio-ecologically resilient than the 'now' of late capitalism. This is what Donna Haraway (2016) means by the phrase 'staying with the trouble', which requires:

> learning to be truly present, that is a vanishing pivot between awful or edenic pasts and apocalyptic or salvific futures, but as moral critters entwined in myriad unfinished configurations of places, times, matters, meanings. (p. 1)

Closely related in function and style are the grounded (Davis 2012) and transgressive (Sargisson 2002) utopia, which variedly call for a utopian transformation of the present from within. The former, exemplified most starkly by historical grassroots movements for social transformation, emphasises a greater 'imaginative awareness' of – and steps towards instantiating – 'neglected or suppressed possibilities for qualitatively better forms of life' within the 'Now' (Davis 2012, p. 136). Exhibiting a Blochean

(1986) emphasis on praxis, these utopian modalities 'of the here and now' (Garforth 2009, p. 14) tend to root their utopian visions in the 'concrete material contradictions of existing states of affairs' (Pepper 2005, p. 7–8). Moreover, these 'zones of otherness' tend prioritise pluralism, diversity, and process over static blueprints (Sargisson 2002). While all utopias harbour transgressive potential, not all can be categorised as truly 'transgressive'; the transgressive utopia similarly arises from 'profound discontent with the political present', yet goes further in rule-breaking, boundary deconstruction, challenging dominant paradigms and creating new conceptual and political spaces (Sargisson 2000, p. 3). Through their 'production and reproduction of spacings and orderings' of radical otherness, they 'shatter taken-for-granted discourses and representational systems' (Garforth 2009, p. 14). Akin to visions espoused by anarchist and post-anarchist thinkers such as Peter Kropotkin (1908), Murray Bookchin (2005), and Gustav Landauer (1978), transgressive and grounded utopias assert that just as important as the need to critique and resist further colonisation of life by the state and corporate capitalism is the need to create counter-spaces within it (Foucault 1986, p. 24; Day 2005, p. 124). Thus, grounded utopias, as exemplified by the alternate traditions of anarchic, anti-authoritarian, and anti-perfectionist utopias (Le Guin 1974, 1985; Huxley 2009), feature complex spatio-temporal forms wherein living, breathing beings engage in prefigurative modes of direct-action in order to realise alternative visions from the past, present, and future in the 'here and now' (pp. 135, 136). Grounded utopias resolutely challenge the purported inescapability of capitalist oppression and exploitation, and the perceived impossibility of eradicating capitalist exploitation within current generations, maintaining instead that already *now* one can – and must – work towards its abolition.

Foucault (Foucault & Miskowiec 1986) epitomises the postmodern turn in utopianism through his concept of the heterotopia,[10] which he

---

10 For Foucault (Foucault & Miskowiec 1986), the ship is heterotopia 'par excellence', as a floating piece of space simultaneously closed in on itself while situated within infinite expanses of ocean (or potentiality), sailing from port to port 'in search of the most precious treasures' ... so much so that, 'in civilizations without boats, dreams dry up, espionage takes the place of adventure, and the police take the place of pirates' (p. 27).

distinguishes from the 'place-lessness' of the classical future-oriented or abstract utopia. For Foucault, the latter run the risk of merely extending the present through projections of existing structural, socioeconomic, etc. conditions. Exhibiting the simultaneous presence and non-presence of utopian ideals, heterotopias, on the other hand, are placeless places. That is, they serve as 'counter-sites' at once within yet, through the gaze set towards other potential worlds, spatial-temporally discontinuous with 'the real' (p. 24). As a result of such developments, much contemporary utopian theory has come to posit utopia not as a dream deferred to a distant time and place (Firth & Robinson 2014, p. 381), but rather as ways of being that can and indeed should be materialised within the present (Davis 2012) via the elucidation and co-creation of micro-exemplars as means of confronting oppressive macro-forces (Firth 2012, p. 31). Thus, some utopian studies scholars have referred to the 'disappearance of the future' (Garforth 2009, p. 12) altogether within some contemporary utopian modalities – in part a reflection of the aforementioned intrusion of widespread socio-ecological disintegration into the 'Now'. Some further contend that utopia's function as transgressive, critical, and subversive of the status quo should be considered in light of how it works to disrupt the present rather than on its reference to an improved future society (Garforth 2009, p. 12). Others (Jameson 1982) have gone so far as to posit that, as prisoners of our cultural and social totality, we are utterly powerless to project or imagine anything outside of it (Jameson 2005, p. xiii). However, as Wells (1913) aptly observes, 'along certain lines and with certain qualifications and limitations a *working knowledge* [emphasis added] of things in the future is a possible and practicable thing' (p. 4). Some orientation towards future potentialities can coincide with the necessary creation of micro-exemplars within the 'Now', and need not equate to a disavowal of diversity and potentiality. What's more, Utopia ought to help guide us in a generally better direction (Suvin 1990, p. 74) while not losing sight of the need to create more liveable spaces within the now. In other words, the 'either/or' approach is antithetical to an effective ethics and politics; preferable is one of 'everything, everywhere, all at once', to allude to the hugely successful recent film of the same name.

Moreover, utopia should not merely seek to embody, explore and articulate an educated desire but rather deploy an educated *hope* that contains the crucial element of praxis – i.e. an active engagement in the transformation of society for enacting better rather than worse futures (Bloch 1986; Levitas 1990; Thompson & Žižek 2014). The post-structural/postmodern turn that disavows closure and prioritises process and criticism of the 'Now' over-imaginative projection (Levitas 2013) must also be weary of falling victim to a 'pathological pluralism' that rejects 'the challenge of literal criticism', thus failing to direct social change towards the better (Levitas 2000, p. 40; Levitas 2003, p. 148). In other words, a complete rejection of closure amounts to 'political evasion' (Levitas 2003, p. 142) and risks a potential descent into nihilistic apathy, which we can ill afford amid the present litany of socio-ecological crises. Negation – while essential and powerful – is insufficient on its own; indeed, in a sense one cannot negate without gesturing towards the possibility of an alternative. The decidedly utopian act of crying 'No!' to the established order (Cathcart 1978) already implies the possibility of another/better 'Not-Yet' (Bloch 1986, p. 307). Necessary emphases on the trials and potential pitfalls of the utopian process, on the inevitable muddying up of pure utopian ideals in the processes of conceptualisation and instantiation should not come at the expense of some vision of *a potential* desired result (Sargisson 2012). For Harvey (2000), intention can be problematic because of its relation to closure around a set of desired ends. However, as Levitas (2003) poignantly reminds us, 'The idea of utopia as spatial play – or social play – may be appealing, but we do not have time to play with it' (p. 150). The litany of mounting Capitalocene crises render critiques of the 'Now', and in particular, concerted steps towards better alternatives, of the essence. While we may be unsure as to precisely which alternate worlds to aim towards, and though conflict and discontinuity around precise details will always persist, what is becoming increasingly apparent is that the present socio-ecological order is wholly untenable.

Ecotopia: The 'Utopia of Radical Environmentalism'

> Where once limpid streams had tapped the snows of mountain flanks and poured their life-giving floods to limitless oceans, where once populous cities, sprawling villages and isolated farms had dotted the planet's surface with the busy hum of activity, now there was lifelessness, death, drought ... Man, the favoured and latest offspring of evolution, had done this ... Deaf to all warnings, heedless of the future, he had denuded the forests, ploughed up the soil, meddled recklessly with the delicate balance of nature. This, in his vanity, he had called the march of civilisation; and an outraged earth struck back.
>
> – Schachner in Ashley (2020, pp. 115–116)

The aforementioned is an excerpt from American author Nathan Schachner's eco-dystopian short story 'The Sterile Planet' published in 1937. Herein, the last remnants of humanity – aristocrats who've barricaded themselves in 'oases' of relative plenty in New York and a new human-like species, 'hordes of the weaker, who, long eras before, had been thrust out from the ever-narrowing oases' (p. 119) that has evolved in the 'deeps' of now evaporated seas, scramble for survival on a dying Earth devoid of topsoil and overrun by desertification. The short story examines timely themes – notably climate apartheid – and, importantly, features a key diagnostic theme prominent in green utopias or 'ecotopias': that of the socio-ecological contradictions of especially industrial-capitalist societies' march towards 'progress'. Green utopianism, or ecotopianism, is a longstanding part of the utopian tradition (Barnhill 2011), with origins in religious and communitarian projects that can be traced as far back as ancient Greek antiquity. A core premise of the green utopian tradition is that there is something fundamentally wrong with the way that industrial-capitalist societies, predicated on endless growth, commodification of Earth others, and profit-oriented exploitation, operate and relate to other terrestrials, and that far more ecologically harmonious attitudes, relations, and modes of subsistence are of the essence (De Geus 1999, p. 210). Thus, of crucial significance in ecotopianism's myriad manifestations is the focus on human relations with nature and other species, particularly, in what ways and to what degrees they can be rendered more egalitarian and less exploitative (Barnhill 2011, p. 130). For many, what they are critiquing and mobilising against is the nightmarish vision outlined above in 'The Sterile Planet'.

Schachner's 'The Sterile Planet' is a notable precursor to the tributaries informing green utopianism, which can be traced especially to the decades following the Second World War now referred to as the 'great acceleration' era (Steffen et al. 2015). Herein, the destructive potentialities of scientific rationality regarding the natural world – i.e. as evinced by the atom bomb – dealt a blow to enlightenment beliefs in benign trajectories of historical and material progress (Linkenbach 2009). A burgeoning ecological consciousness thus emerged from the late 1960s and early 1970s onwards (more in the following chapter), underpinned by wider emerging trends in u(ecotopian) thought associated with increasingly vociferous 'limits to growth' discourses (Meadows et al. 1972) and the 'deep ecology' movement in eco-philosophy (Naess 2005). These were reactions against increasingly volatile modes of capitalist expansion and accumulation. The new ecological consciousness stemming therefrom harboured anxieties over a planet increasingly imperilled by worsening human transgressions of planetary boundaries via expanding populations and consequent socioeconomic activity.[11] They loudly proclaimed the impossibility of endless growth on a finite earth system (Schumacher 1973; Garforth 2018). Unsurprisingly the aforementioned developments spurred the production of scientific eco-dystopias such as Margaret Atwood's Maddaddam Trilogy (2014) and ecotopias (Linkenbach 2009, pp. 5, 10) such Callenbach's *Ecotopia* (1975), Ursula Le Guin's *Always Coming Home* (1985), Starhawk's *The Fifth Sacred Thing* (1993), Kim Stanley Robinson's *Pacific Edge* (1995) and Aldous Huxley's *Island* (2009). Whilst eco-dystopian works warn of the potentially disastrous socio-ecological consequences of endless material expansion and

---

11   Long before this, however, Alexander Bogdanov's Bolshevik utopian novel *Red* Star, first published in 1908, features a communist utopia on Mars grappling with similar concerns over transgressing planetary boundaries; ultimately as the Martians put further strain on their forests, food and energy supplies, and refusing to consider population control measures, they begin to debate the need to colonise either Earth or Venus (Bogdanov 1984, p. 79).

nature exploitation, ecotopias creatively depict worlds devoid of rampant ecological despoliation.

Similar anxieties over the perceived dangers of transgressing ecological limits through virtually unhindered socioeconomic expansion also strongly influenced the green movements in Europe and North America (Linkenbach 2009, p. 9). Hence, Pepper's (2007) reference to ecotopia as the 'utopia of radical environmentalism' (p. 289). As Dobson (2003) poignantly observes, the common argument detected amongst some reformist environmental discourses that the natural world and other species must be preserved either largely or exclusively for the continuity of *human life* is not an ecotopian argument. In failing to critique the structural underpinnings of the violent and reductive treatment of other species, and in failing to value the other as such, these approaches don't entail the fundamental transformation of human values and modes of relationality with regards to other terrestrials. What the limits discourse does often lack is the imaginative component integral to utopianism, namely the projection of alternative sensibilities and modes of being. Ecotopias urge that societal and ecological wellbeing can only be achieved by respecting the biogeophysical constraints imposed by the wider ecosystems in which we're embedded, limits that render boundless industrial/growth-oriented socioeconomic systems ecologically nonviable as well as ethically problematic. Ecotopias aim to deconstruct the very notion of a separation between the 'social' and the 'natural', emphasising that social and ecological justice are two sides of the same liberation coin. Moreover, a concern with ecological sustainability and respecting wider biospheric dynamics doesn't mean that ecotopias are static or devoid of fluidity and change; observance of overarching biogeophysical principles can and often does coincide with dynamic and fluid socio-ecological interactions in ecotopias (Dobson 2003). 'Freedom' in Ecotopia is instead qualitatively redefined along ecological lines and in relation to our terrestrial counterparts. The 'good life', no longer synonymous with the boundless acquisition of material goods and wealth by and for humans alone, is rather re-envisioned as that which promotes a mutual, multispecies flourishing.

The above can be seen to varying degrees not only in ecotopian literary works and social movements (Chapters 4 and 5) but also in the thousands of green intentional communities that have proliferated globally (Federation for Intentional Community 2022)[12] in the form of housing co-operatives, communes, ecovillages and co-housing communities (Sargisson 2012, p. 131; Firth 2019). Since the 1980s, and especially since the mid-2010s, ecovillages have emerged as one of the fastest-growing forms of intentional community (Sargisson 2012, p. 131). There are over 2,000 in the US alone, such as Twin Oaks, an intentional community founded in Louisa, Virginia in 1967 by people inspired by B. F. Skinner's utopian fiction Walden Two (Sargisson & Sargent 2017, p. 14) and which is now a significant ecovillage. Lucy Sargisson (2007, 2009) and Lyman Tower Sargent (Sargisson & Sargent 2004) have done extensive work elucidating these lived exemplars of Ecotopia in the 'Now', with the UK and New Zealand among the prime case studies. Sargisson and Sargent (2004) define green intentional communities (GICs) as collectives formed by people – at least five adults and their children (Sargent 2004) – who choose to live together over a shared concern over environmental issues and/or ethics, so that they might live more closely with the natural world and help preserve it (p. 113). In other words, they aren't merely utopian experiments in new socio-political and economic arrangements; Importantly, they foreground ecological resilience and self-transformations for realising new ontological and ethical relationships with the land and other terrestrials (Sargisson 2007). They are, in effect, ecotopias in-concreto. In New Zealand, as Sargisson and Sargent (2004) point out, Māori culture and cosmology serve as an important background influence underlying the prevalence, nature and aims of GICs, namely the divine links between humans and the land, and customs of communal land tenure (p. 115; also see Chapter 6).

Though they generally seek to radically reconfigure human-nature relations, not all green utopias are created equal. Technocratic Sci-Fi utopias influenced by modern ideals of progress emphasise technological development and material abundance (within which the Venus Project and H. G. Wells' utopian writings might be situated) as key features of better worlds; 'deep green' utopias, on the other hand, tend to approach technology with greater scepticism, or rather, are critical of an over-reliance on technocratic

[12] https://www.ic.org/.

solutions to complex socio-ecological issues. Instead, they call for more modest modes of life predicated on proximity to – and more harmonious interactions with- nature and other species (i.e. Huxley's *Island*) (De Geus 1999; Garforth 2005). In the deep green utopia, biospheric flourishing and notions of individual and societal wellbeing are decoupled from ceaseless material progress and acquisition, and reconceptualised so as to place greater emphasis on nurturing meaningful relationships for socio-ecological wellbeing (De Geus 1999). The deep green utopia, as opposed to its more reformist or technocratic variants, advocates a radical break with the present by embodying and gesturing towards a radical ecological consciousness and direct-democratic political modalities. It is within the latter category that radical environmental activists (REAs) and the literary ecotopias examined in the following chapter might be situated. Moreover, unlike (some) utopian socialists and technocratic utopians who have tended to believe in social perfectibility through an almost dogmatic faith in scientific advancement, 'the empathies of today's radical environmentalists are frequently quite different, often resonating with postmodern skepticism about such ideas' (Pepper 2007, p. 291). In their seemingly anti-hierarchical and post-anthropocentric worldviews, they deviate quite considerably from more anthropocentric and technocratic utopias such as Wells' *Men Like Gods* (1923). The latter depicts a 'utopian' society which, in an ultimate display of anthropocentric hubris, has deployed techno-scientific progress towards a 'systematic extermination of tiresome and mischievous species' (Wells 1923, p. 76) like mosquitos, and otherwise wholly subjugated nature in service of humanity[13] (Alberro 2022). Though REAs' general opposition to the

---

13 In J. D. Beresford's cautionary short story 'The Man Who Hated Flies' (in Ashley 2020) a scientists bent on eradicating flies from the world discovers a pathogen that proves lethal to them, only to set off a chain reaction of ecological catastrophes after the pathogen begins to infect other insects including pollinators who play a vital role in food production (p. 67). As birds and other species dependent on flies disappeared in tow, 'something of music had gone from the earth. The world was stiller than of old, less beautiful, moribund. There was less colour, less variety, less vitality' (p. 69). A prescient precursor to Goulson's urgent *Silent Earth: Averting the Insect Apocalypse* (2021b). Indeed, Wells also wasn't unaware of the potentially disastrous ecological consequences in the form of trophic cascades that could result from the elimination of certain species, as a character in *Men Like Gods* subsequently remarks on the paucity of birds and other fauna as a result of the extermination of insects.

Western, industrial-capitalist 'megamachine' (Bahro 1994, p. 34) and its increasing infringements on the earth system does not equate to a wholesale repudiation of technology and the scientific enterprise per se. Rather, as discussed, what they are critical of is the hubristic and deleterious use of technology in service of the exploitation and subordination of Earth others, as well as an excessive reliance upon technocratic approaches for ameliorating our socio-ecological woes at the expense of more fundamental structural, socioeconomic, and cultural transformations.

Theoretical Ecotopias: Critical Posthumanism

In Chapter 2 we explored the historical and theoretical tributaries of human supremacy, a way of inhabiting and knowing the world and others that has subjected myriads of our terrestrial kin to a litany of dystopian horrors. Utopianism shares a close affinity with critical theory (Marcuse 2013) more generally in its aims to critique and deconstruct antagonistic dualisms, unequal, oppressive and exploitative structures and modes of relationality (Braidotti 2016; Ulmer 2017). Critical-posthuman theories direct this critical gaze towards human relations with other species; in this sense it can be framed as a theoretical form of (eco)utopianism. It annunciates ways of relating to other species that are more egalitarian and respectful, as desirable alternatives to the aforementioned system of human supremacy. As such it has many affinities with indigenous cosmologies and ethico-political modalities, as we will explore further in Chapter 6. Though stemming from a multiplicity of theoretical tributaries and traditions, critical posthumanism can be associated with certain core characteristics. In particular I draw largely on Braidotti's (2013) approach which posits the posthuman dimension of post-anthropocentrism (Ferrando 2016) as a 'deconstructive move' that seeks to dismantle the foundations (conceptual and otherwise) of 'species supremacy', whilst inflicting a blow to 'any lingering notion of human nature, *Anthropos* and *bios*, as categorically distinct from the life of animals and non-humans, or *zoe*' (p. 65). Critical posthumanism avows a non-hierarchical, non-dualistic and relational ontology that recognises a 'deep zoe-egalitarianism' between humans and other species (Braidotti 2013, p. 71). In this it also shares an affinity with feminist, queer and postcolonial theoretical tributaries and forms of political praxis, which variedly seek to deconstruct antagonistic, exploitative and reductive orientations historically marginalised human 'others'. The self and the 'we' are thus seen not as neither unitary or universal but rather as complex 'nomadic assemblages' that are always 'relational, transversal and affirmative' (Braidotti 2016, p. 16). Many strands of posthumanism express a fundamental 'dissatisfaction and/or rejection of the "two central tenets of humanism": namely, the belief that humans are the center of the world (i.e. anthropocentrism) and that, as superior rulers of existence, we have the right to subdue, exploit, and/or otherwise reduce unruly "others" to the status of object' (Taylor 2012, p. 37). Viewed from a posthuman perspective, the notion of the human as an autonomous and self-contained entity is no more; humans, like all other entities, are thoroughly hybrid beings. The related and more precise term 'post-anthropocentrism' similarly refers to orientations that are concerned with critiquing and deconstructing human-exceptionalist worldviews as well as species hierarchies (Braidotti 2016; Ferrando 2016).

Posthumanism as it is referred to throughout does not entail the view that humans are insignificant within the wider universe; rather, especially in its critical strands, it seeks to bring about an ethical and ontological re-alignment of the human vis-à-vis the other-than-human world. Post-humanisms more generally reject such longstanding assumptions (Heidegger 1995) as that human ways of knowing and being in the world are *essentially* different from and superior to those of other-than-human entities (Plumwood 2002). Importantly, critical posthumanism is to be distinguished from certain strands within the wider and multifarious 'posthuman' tradition such as transhumanism. The latter celebrates the continuation or extension of the human via biotechnological advancements, and generally evinces an enthusiasm over the increasing technification of contemporary societies (Botz-Bornstein 2012; Ferrando 2016). Such approaches avow an 'extension' of the human as desirable, and are concerned less with attempts to respect and relate more ethically to other species as our fellows. Of interest for the purposes of the current research – and as will be elucidated in subsequent chapters – is the degree to which REAs (and canonical ecotopian literary texts) adopt critical-posthumanist approaches to the other-than-human world and in their delineations of ecotopian alternatives. Particular attention is given throughout not only to the posthuman thought of Bruno Latour (2004, 2011, 2017, 2018) and Jacques Derrida (1995a; Derrida & Wills 2002) but also Val Plumwood (2002, 2008), whose philosophical animism (Rose 2013) and explorations of the politics and ethics of dwelling in multispecies ecological communities are deemed especially pertinent in relation to subsequent discussions. Such expansive and radically 'other' approaches to human relations with other terrestrials offered by critical-posthuman thought, which seek to deconstruct oppressive anthropogenic hierarchies, harbour considerably transformative potential and thus remain of interest for discussions surrounding the ecotopian 'Not-Yet'.

From Hierarchies to Assemblages: Latour's Irreductionist Ontology

Within the tradition of posthumanist scholarship, Bruno Latour is one figure who stands out (in my view) in terms of his thorough and far-reaching critique of the structural and onto-epistemological foundations of anthropocentrism (2004, 2011, 2017, 2018). In Latour's non-hierarchical ontology – which operates on a plane of immanence – existence, other beings, objects, etc., simply *are*, with no hidden essence underlying reality or structures that shape them and their various doings (Harman 2009). Therefore, the Aristotelian-Leibniz schism between matter and form (Harman 2009, p. 106), wherein 'primary matter' serves as the essential and enduring corporeal substance that underlies the 'secondary matter' of concrete forms, no longer holds.[14] In rejecting the notion of

14 Leibniz observes that, 'I agree, sir, that there are only machines (that are often animated) in all of corporeal nature, but I do not agree that there are only aggregates of substances; and if there are aggregates of substances, there must also be *true substances from which all the aggregates result*' (cited in Garber 1997, p. 334). In other words, aggregates or assemblages must ultimately be grounded in some unified substance in order for them to be 'real', a notion that is rejected by Latour's flat ontology.

essence and therefore of entities possessing an inner substance as distinct from fluctuating surface attributes, actants – humans, hurricanes, cyanobacteria, COVID-19, the neoliberal paradigm, NATO – are (re)conceived as concrete and specific events in space and time. They are always completely present in the world, though in a continuous state of flux and becoming, hence their radical uncertainty (Harman 2009). The very nature of 'external reality' is surprise, rather than the 'simple "being there" of objective matters of fact' (Latour 2004, p. 79). All actants regardless of size or complexity are posited as concrete entities that affect and are affected by others on an equal playing field, wherein none are granted special ontological status. Reality is thus (re)conceived as a vast, pluralistic, horizontal network of ever-shifting alliances between various actants who co-produce one another, gain as well as lose power[15] through their alliances, and indeed *are* their alliances. The world emerges as a 'series of negotiations between a motley of armada forces, humans among them, and such a world cannot be divided cleanly between two pre-existent poles called "Nature" and "Society"' (Harman 2009, p. 13). Interaction and negotiation are central here, for actants must continuously and actively negotiate with others in their myriad strivings, in congenial alliances as well as clashes of wills with recalcitrant actants who often resist to varying degrees in their own strivings. The relative successes and failures of such negotiations are never established a priori but must be actively maintained, hence why the 'future-to-come' is no given but must be actively and continuously co-constructed (Harman 2009, p. 19).

Latour is relevant to discussions around boundary construction and deconstruction, and discussions around ecotopian politics, as he dissolves human/animal, subject/object, and even animate/inanimate dualisms through his concept of an actant. An actant is essentially anything that has goals, can alter the course of events outside of itself, and alter its course in response to changing circumstances (Latour 2005, p. 72). Similarly, he undermines the very notion that there is a 'social world' as distinct from the 'natural world', and likewise a 'social' science as distinct from the

---

[15] That is, power not in the Machiavellian sense but as in the capacity to do and to make things happen.

'natural' sciences. Thus, attempts at reconciling 'Nature' and 'Society' are futile because the supposed dualism between the two never was in the first place[16] (Harman 2009, p. 58). Agency, diffuse and exhibited by actants of myriad kinds, rather than an attribute of conscious intentionality and reflexivity, is therefore not something that only humans or even most other-than-human animals possess. Rather, it is a property of dynamic matter itself (Bennett 2010), yielding a world full of entities with the power to make things happen, and to either thwart, evade or aid the ventures of others. Such a view would then somewhat complicate assignations of culpability for Capitalocene crises, for as we are always in alliances with other actors, not even humans – influential though many of us have become as of late – act alone but are always 'coextensive' with myriad other actants such as carbon, glaciers, hurricanes, and plastics. The most radical contribution of Latour's Actor-Network Theory (ANT) (2005) and wider onto-epistemological framework is thus the significant strides it makes in dismantling the ontological foundations of human supremacy and the logic of dualism (Plumwood 2002) more generally. Humans, no longer situated aloft and disconnected from the rest of existence and other entities, are firmly re-situated within the vast multispecies assemblages that constitute reality, moving and striving alongside and in conjunction with other actants. Indeed, 'without the nonhuman, humans would not last a minute' (Latour 2004, p. 91). In subsequent chapters I variably draw on Latour for aiding discussions of the nature of ecotopian collectives, specifically deliberations around how we ought to relate to our co-terrestrials, as well as the extent to which dualistic constructions of the human relations with other terrestrials are dismantled in contemporary ecotopias.

16 Latour's (1991) notion of time is equally non-modern; the Western concept of a linear progression from past to future is eschewed in favour of one of 'spirals and reversals' akin to shifts in inter-actant networks, wherein one's genetic code, nation states, technological instruments, and the like are motley mixtures of different time periods (Harman 2009, p. 68). Events in a given year can both be said to precede subsequent years while also having portions that are produced only *retrospectively*. In other words, 'the more sensitive the chromatographer, the more realities abound' (Latour 2004, p. 85).

Towards Multispecies Justice

Another promising development associated with a critical-posthuman turn that can serve as a useful lens for informing analyses and explorations of terrestrial ecotopianism, which crucially values and foregrounds our terrestrial kin, is the 'multispecies justice' (MSJ) (Haraway 2013) activist and theoretical movement. As espoused by sociologist Danielle Celermajer, leader of the Multispecies Justice Project, and colleagues, a multispecies justice (MSJ) lens:

> expands the existing frame of climate justice by repositioning justice to encompass *all beings as quintessentially relational* [emphasis added] ... This MSJ lens makes it possible to transcend liberal individualist and anthropocentric notions of justice, in two fundamental ways: first by enlarging the range of obligations and duties is-à-vis all those hose flourishing is undermined (Celermajer, Schlosberg, et al., 2020); and second by shifting the focus and subject of justice from the individual and exceptional human being to a wide range of living and non-living entities. (Tschakert et al. 2021, p. 4)

The MSJ movement emerged out of, and continues the vital focus on, the foundational concept of intersectionality (Crenshaw 2017) that traces environmental exploitation as fundamentally and inextricably intertwined with issues of social injustice (Tuana 2019; Tschakert et al. 2021). Environmental and climate justice movements and discourses have been hugely consequential in their elucidation of how colonial legacies, structural racism and other axes of oppression continue to disproportionately impact and render further vulnerable marginalised groups (Schlosberg et al. 2014). However, as denoted above, even in the important framing of 'climate justice', the subject of justice rarely if ever expands beyond the human. The MSJ movement builds upon the essential work of illuminating and resisting the complex interactions between racism, environmental exploitation, and unequal exposures to ecological harms amongst vulnerable human populations, whilst expanding such concerns to other terrestrials as subjects of justice (Tschakert et al. 2021). As such, MSJ widens the realms of possibility when it comes to imagining ecologically resilient and just futures, whilst mobilising and fighting for a much more expansive understanding of justice and community. As with critical posthumanism,

A Better World is Reality

an MSJ framing avows an inclusive and relational ontology, ethics, and politics firmly rooted in the vast relational webs of coexistence that envelop all terrestrials. It also seeks to embrace interdependencies as well as heterogeneities amongst humans and other species.

One promising example of MSJ is the global and decidedly ecotopian 'ecocide' movement, an agglomeration of legal scholars and activists aiming to make ecocide – or systematic ecological destruction – an international crime within the jurisdiction of the International Criminal Court (ICC) (Alberro & Daniele 2021). As I've discussed in previous chapters, radically reconfiguring our ideas about justice, politics, community and 'the good life' is a major ethical imperative of our times; in addition to this, it is an existential imperative, for as Haraway (2018a) reminds us, 'there can be no environmental justice or ecological reworlding without multispecies environmental justice' (p. 102). Not only justice – there can be no habitable terra, no future – whether liveable or nightmarish – no 'us' without our terrestrial kin. However, it's also important to remain critically aware of the ambiguity, elusiveness and at times impossibility of justice, of the potential limits and pitfalls of representation. In the creatures and communities being represented in our strivings for MSJ, 'words often do fail us" (Chao et al. 2022, p. 5). At its core, MSJ can be conceptualised as a 'promise' to do better with our words, and to 'tend to our fellow terrestrials and the communities that we love' (p. 6).

Terrestrial and Post-Terrestrial Strivings!

Not all utopian visions and strivings are created equal; the danger lies in exclusive utopias – such as the free-market utopia of neoliberal capitalism – which reserve 'the better' for a select minority (Sargent 2006, p. 12). Similarly, as Suvin (1990) warns, there's a vital distinction to be made between the 'pseudo-novum' of commodification that dominates the terrain of the new within late-stage capitalism (Moylan 2021), and a radical (utopian) novum that intervenes in and works to dismantle oppressive and exploitative systems and relations (Suvin 1988;1990). This is why having some sense of where we want to go collectively, and remaining critically self-reflexive of the content and direction of our visions, is so important. As Murtola (2018) observes: 'How the future is imagined matters. It matters for planning and it matters for action … Right now, billions are being poured into competing visions of capitalism in space' (p. 2, 7). Post-terrestrial escapist fantasies are increasingly pedalled by the global techno-capitalist elite, as seen in Elon Musk and Jeff Bezos' plans to flee to and colonise extra-terrestrial worlds. Musk's SpaceX programme seeks to make the colonisation of Mars a possibility within the next decade (for *only* $200,000 per person[17]) in order to preserve *human* civilisation in the event of a third world war. His other dream is to see humanity become a 'multi-planet species' (Williamson 2016; Musk 2018). Yet their exclusive utopian visions banish a multitude of human and other-than-human terrestrials via a form of 'climate apartheid' wherein the wealthy elite effectively buy their way out of ecological breakdown. Moreover, Bezos and Musk's post-terrestrial utopias see in the horizon of possibility surrounding the real only novel opportunities for extending capital's exploitative, expansionary and exterminatory necropolitics to distant times and places, just as the 'New World' served as the next frontier of exploitation for European colonial powers (Mbembe 20008). Their novum is one of redoubled commodification and colonial conquest. In his recent work *After Lockdown: A Metamorphosis* (2021), Latour refers to such 'mad' escapist desires as a dangerous form of secularised religion which has merged 'God and the 'Dollar', legitimising a profligate destruction of resources whilst leaving scores behind to fend for themselves (p. 57).

In stark contrast, the utopianism exhibited by many of today's progressive social and environmental movements is, literally and metaphorically, worlds apart. Fuelled by the conviction that another Terra is possible, movements like the socialist Zapatistas, Occupy Wall Street, and more recently the Sunrise Movement and Extinction Rebellion, all began by

---

17   Hence dismissals of such programmes as inherently elitist. Linda Billings (2019), a social scientist and consultant to NASA's Astrobiology Program, poignantly enquires: 'How many poverty-stricken Bangladeshis, how many sub-Saharan Africans, how many permanently displaced Syrian refugees, how many disabled and unemployable workers could come up with $200,000 – or $2,000, for that matter – to move to another planet and start a new life?' (p. 45).

crying a resounding 'No!' to austerity and bank bailouts, wars, fracking, pipelines and oil subsidies, to the theft of land from local and indigenous communities, the privatisation of water and other vital resources, and to the increasing concentration of wealth in the hands of a parasitic global minority. Rather than casting their sights to distant times and places in search of new colonial frontiers, these movements seek to build a better world for all in the here and now. They and especially the radical environmental movements that I'll be exploring in subsequent chapters evince what I term a 'terrestrial utopianism', following Latour's call for a utopian politics rooted in the 'here below', rather than aforementioned modalities that seek to 'beam us into the beyond' (2017, 2018, 2021). Building on Moylan's (1986, 2021) critical and Davis's (2012) grounded utopias, Terrestrial utopias are praxis-oriented (Bloch 1986), intent on effecting a total liberation from oppression and exploitation through allied networks of collective resistance. Terrestrial utopias are thoroughly rooted in terra (Latour 2017, 2018) and especially foreground more respectful relationality with our terrestrial kin. As such they draw heavily on critical-posthuman tributaries and longstanding indigenous cosmologies in avowing a thoroughly relational ontology, ethics and politics which sees other-than-human terrestrials as valued and agentic beings with whom we always intra-are. Terrestrial utopias' grounded disruptions of the 'Now' negate the politically fatalistic, anti-utopian ideology and commodified socioeconomic relations associated with advanced capitalism, which offers the 'good life' to the highest bidders (Sargent 2006). The good life must be re-examined and posited with, by and for, all terrans.

A Continual Striving

> ... the present pain of living in the world is perhaps in some ways unprecedented. I write in the night, although it is daytime ... People everywhere – under very different conditions – are asking themselves – where are we? The question is historical not geographical. What are we living through? Where are we being taken? What have we lost? How to continue without a plausible vision of the future?
>
> Berger (2007, pp. 35, 6).

Such poignant queries must be asked in a temporal as well as spatial sense – the latter in relation to which terrestrials we share a given space with, and how we can incorporate them and their concerns into ecotopian collectives. Utopias in their myriad forms are the embodiment of productive negation and hope, 'great refusals' (Breines 1989) of the 'Now' born of concrete experiences of lack during an interregnum where the existing status quo, its dominant norms and institutions is dying, and new modalities are struggling to burst forth (Marcuse 1969; Gramsci et al. 1971; Thompson & Žižek 2014, p. 2). Yet, it is precisely when the internal structure and cohesion of the capitalist system starts to disintegrate, via socioeconomic, political and ecological upheavals, that radical alternatives most thrive (Marcuse 1969, p. 82). Utopianism's imaginative component seeks the renewal of society's 'cell tissue' (Buber 1996); that is, a (re)construction of its essential building blocks – the individuals and modes of relationality that comprise the social (or socio-ecological) world – so as to transition from capitalism's 'hollowing out' and erosion of diverse socio-ecological assemblages towards life-affirming and all-embracing collaborations (p. 14). The utopian 'now' as the site of revolutionary agency – embodied by REAs, intentional communities, and transgressive literary utopias – interrupts traditional homogenous empty time that is a defining feature of commodity capitalism. In so doing, utopias generate a temporal breach wherein alliances with other moments and spaces in time cut 'through the fabric of this world and reveal[s] world after world of possibilities and forms' (Starhawk 1993, p. 472). Moreover, utopias work best when they cast their critical gaze not only on hegemonic definitions of reality that constrain possibility, but also on themselves, so that they avoid stasis, ossification and parochial isolation. The multiple socio-ecological dislocations of the present demand a move away from the prioritisation of utopianism's negative or denunciatory function and towards its powerful capacity for annunciation (Moylan 2021, p. 113). There is no time to spare; we can no longer afford the luxury of mere critique. This recalls Levitas's (2003;2007) important observation that, indeed utopian theory and praxis must maintain a degree of open-endedness and fluidity, yet not so open-ended that we become politically evasive and thereby fail to direct social change towards the better. It's a complex yet crucial balance to maintain: engaging in a modality of utopianism that avows the dynamic and always-unfolding nature of reality, whilst not losing sight of the urgency of transformative political action for bringing about better worlds.

The ecotopian manifestations that I'll be exploring in the following chapters variedly gesture towards better socio-ecological alternatives in their inclusive orientations and modes of relationality, demonstrating through their very existence that other modes of being are possible (Davis 2012). However, the extant and looming climate catastrophe, biological annihilation, and biogeophysical transformations will likely engender profound transformations in the utopist's very capacity to imagine the future. Contemporary utopian and ecotopian modalities are already living and mobilising within the 'end times' of environmental disintegration. This makes it all the more vital that we become 'present again to the situation of terrestrial rootedness' (Latour 2017, p. 212) by learning to feel our responsibilities to, and find ways to live more ethically and fairly with, our co-terrestrials. For the purposes of this book, the interest is especially in how contemporary ecotopian manifestations frame and feature long-excluded other-than-human terrestrials in their denunciative and imaginative practices and projections of better worlds. Some version of the good place might not be 'no place' after all, but rather reside within life-affirming political mobilisations actively resisting the 'now' in service of a 'to-come' that is more conducive to a multispecies flourishing.

> We should hold a steadfast orientation towards the open ocean of possibility that surrounds the actual and that is so immeasurably larger than the actuality. True, terrors lurk in that ocean: but those terrors are primarily and centrally not (as the utopophobes want to persuade us) the terrors of the not-yet-existing, but on the contrary simple extrapolations of the existing actuality of war, hunger, degradation, and exploitation of people and planets. (Suvin 1990, p. 81)

CHAPTER 4

Literary Ecotopias

Literary utopias counter the present by foregrounding and offering vivid textual details of alternative social values and political practices that are otherwise rendered invisible by the prevailing common sense (Moylan 2021, p. 108). My focus in the subsequent discussion of selected ecotopian texts is less on style and form, important though these are (for instance in foregrounding details about terrestrial life and agency typically relegated to the background in other texts) in relation to a given works' content. Crucial for the ecotopian literary text's transformative capacity is its 'estranged method' – i.e. the cognitive estrangement spurred by the role of the visitor, for instance (see Chapter 3), which allows the reader to develop critical perspectives on their own society (that under critique) by way of its juxtaposition with the text's imaginative representation of a better world (Moylan 2021). This productive disjuncture is the crucial stepping stone that paves the way for the enactment of the better. Rather, I pay particular attention to content, through a critical-posthuman approach to one of literary ecocriticism's key problematics: how the other-than-human interacts with and is represented by human cultures (Heise 2013). Specifically, my interest is in how the texts below depict and (re)configure relations between humans and other terrestrials in their ecotopian critiques and projections of sustainable worlds. Ecocriticism shares the utopian imperative of reshaping human-nature relations along more harmonious trajectories (Heise 2006). Material ecocriticism further draws on critical-posthuman tributaries in order to shed light on the myriad narrative agencies with which we share the world, wherein storytelling is far from an exclusively human practice. Thus, a material-ecocritical lens can help us reflect on 'the many ways earthly forces and beings can speak', thereby compelling us to 'rethink our coexistence and coevolution in the story of the earth itself' (Oppermann 2014, p. 89).

In *The Great Derangement: Climate Change and the Unthinkable* (2018), Amitav Ghosh points to the seeming reluctance of literature to effectively deal with the immediate urgency and unpredictability of climate change, a predicament that he partly relates to the enduring legacy of human supremacy in the West. Ghosh denotes that the (Western) literary imagination thus remains 'radically centred on the human' (2018, p. 114). A common emphasis within the (deep) green utopian canon is on qualitative reconfigurations of interspecies relations, specifically a focus on forging meaningful interpersonal relations over material acquisition and expansion (De Geus 1999; Garforth 2018). However, as I'll discuss below, the extent to which other species occupy the foreground of ecotopian texts as valued and agentic beings alongside human characters varies considerably. While substantially more harmonious multispecies relations are depicted in many of the ecotopian texts under analysis in this chapter, the singular and embodied experiences of other species are seldom explored in detail.

The texts I examine further below are: Ernest Callenbach's *Ecotopia* (1975), Aldous Huxley's *Island* ([1962] 2009), Ursula Le Guin's *Always Coming Home* (1985), Kim Stanley Robinson's *Pacific Edge* (1990), Starhawk's *The Fifth Sacred Thing* (1993) and Rupprecht et al.'s (2021) *Multispecies Cities: Solarpunk Urban Futures*. These particular texts were selected because of their specific engagement with human relations with other species in their critiques of dominant systems and imaginative projections of more socio-ecologically sustainable futures, hence why many are regarded as canonical 'ecotopian' texts (Mathisen 2001). They also variedly fall within the genre of the 'critical utopia' in their attempts to subvert linear conceptions of time, fragmented narrative structures, and unresolved dilemmas and tensions. The selected texts to varying degrees exhibit the green-anarchic and 'post-industrial' utopianism advocated by the likes of André Gorz (1985) and Rudolf Bahro (1984) (Frankel 1987) in their calls for decentralised, anti-hierarchical, relatively small-scale, interconnected and egalitarian eco-communities. Huxley's *Island* is a curious outlier in the selection of works because it was originally published a few years before the limits-to-growth debates that sparked the emergence of subsequent green utopias as well as radical environmental movements;

it was selected because of its prescient concerns over the deleterious ecological consequences stemming from unchecked industrialisation and socioeconomic expansion (Mathisen 2001), as evident in its depiction of a decentralised, self-reliant ecotopian society observant of ecological 'limits' to the human enterprise. These are, as will be discussed in the following chapter, many of the same concerns and desires shared by contemporary REAs. Indeed, Huxley's novel served as an inspiration for subsequent ecotopian texts including Callenbach's *Ecotopia* (1975). *Ecotopia*, distinct in style and content from critical ecotopian works such as *Always Coming Home*, is the classic work that lent the ecotopian literary tradition its name (Moylan 1986; Garforth 2018), thus serving as a useful point of comparison. Lastly, *Multispecies Cities* is a wonderfully pertinent recent anthology that features an array of stories attempting to grapple with making kin in a damaged world. Whilst not a traditionally ecotopian novel in terms of structure and style, *Multispecies Cities* features palpable ecotopian themes and visions of the possibilities for multispecies flourishing amidst and beyond the Capitalocene.

Islands in the Stream

> Only when we get it into our collective head that the basic problem confronting twentieth-century man is an ecological problem will our politics improve and become realistic … Do we propose to live on this planet in symbiotic harmony with our environment? Or, preferring to be wantonly stupid, shall we choose to live like murderous and suicidal parasites that kill their host and so destroy themselves?
>
> – Huxley (1963, p. 6)

As has been previously noted, the utopian (and dystopian) imaginary in its myriad incarnations proliferates during times of precariousness and upheaval. Huxley's *Island*, originally published in 1962, is the utopian counterpart to his arguably more famous dystopian novel *Brave New World* (1932), published about thirty years earlier (Schermer 2007). The story is centred around a small island called Pala in the Pacific, where a utopian society has flourished for 120 years. The novel's key protagonist Will Farnaby, who plays the role of the visitor, is a journalist who has shipwrecked there, although it is revealed that he is actually an underground agent for an oil giant sent to work out a deal with Pala's government in order to gain access to the island's substantial oil reserves. *Island* was born during the height of the geopolitically turbulent Cold War era and the increasing ecological perturbations of the post-1945 'great acceleration era' of mass consumption and production. Like Rachel Carson's paradigm-shifting work *Silent Spring* published during the same year, reflection of the burgeoning ecological zeitgeist of the time, *Island* similarly echoes concerns over the mounting socio-ecological deficiencies stemming from rampant industrialisation (whether under capitalist, communist, or any other hegemonic system) and over-consumption. *Island* was published nearly a decade prior to the early-mid 1970s 'limits to growth' debates in green political discourse that similarly influenced the emergence of radical environmental groups (Mathisen 2001).

Huxley harboured concerns over the social, political and ecological ramifications of unchecked human population expansion which he expounds upon in his essay *The Politics of Ecology: The Question of Survival* (1963),[1] and which feature prominently in *Island*. In Pala, in light of free and wide access to contraceptives, the lack of religious custom discouraging their use, a cultural ethos of living within the carrying capacity of the island, denizens never have more than three children and 'most stop at 2' (p. 80). The socioeconomic and political system in Pala is summed up, somewhat problematically, along the following lines: 'Electricity minus heavy industry plus birth control equals democracy and plenty' (p. 144). An important though not always explicit critique in many ecotopian literary diagnostic narratives is profligacy amongst the elite few at the expense of extreme deprivation for the multitudes, a reflection of the close relationship between socioeconomic inequality and environmental degradation. In the following scene from *Island*, the novel's anti-utopian character, The Rani, notes:

---

1   This essay contains some deeply problematic and colonial assumptions such as that under-developed nations are too beset by poverty, ignorance and simple-mindedness to exercise control over their 'uncontrolled breeding' (p. 3).

Bahu (Rendang ambassador, an area on the periphery of the utopian society of Pala) is the Last of the Aristocrats. You should see his country palace! Like the Arabian Nights! One claps one's hands – and instantly there are six servants ready to do one's bidding. One has a birthday – and there is a fête nocturne in the gardens. Music, refreshments, dancing girls; two hundred retainers carrying torches. The life of Haroun al Rashid, but with plumbing. (p. 49)

At which point Will Farnaby sardonically retorts: "'It sounds quite delightful,'" while remembering the 'villages through which he had passed in Colonel Dipa's white Mercedes – the wattled huts, the garbage, the children with opthalmia, the skeleton dogs, the women bent double under enormous loads' (p. 49). Egalitarian social relations amongst human collectives in ecotopia are framed as not only as ethically desirable but also essential for ensuring less destructive relations with the other-than-human world. As such, Palanese society is a federation of small-scale self-governing units, an ideal alternative to the rampant consumerism and industrialisation seen within both capitalism and communism as large, totalising systems (Mathisen 2001). It also has no army, a fact that Colonel Dipa, another anti-utopian figure, wishes to overturn, whilst heavily industrialising Pala (p. 47). Indeed, the brutality and wanton waste of war is a theme that often features in many utopian and ecotopian texts. In an explicit attempt to move away from the ravages of capitalism, particularly extreme inequality as alluded to above, Pala has implemented an economic system predicated on sufficiency and cooperation rather than ceaseless growth and competition; income and wealth inequalities are kept in check by the mandate that no denizen is permitted to become 'more than four or five times as rich as the average' (p. 146).

In *The Politics of Ecology* (1963), Huxley asserts that a fundamental error in our dealings with other terrestrials is our hubristic assumption that we are not members of earth's ecological community, which in turn allows us to 'throw our weight around like gods' (p. 6). Though he believed that other animals have no souls, he didn't think that this ought to translate into our treatment of them as mere things (p. 6). Rather, we should treat them as 'parts of a vast living organism', and apply the golden rule of 'do unto others as you would be done by' to our dealings with them and the wider natural world. Such is the ethic that informs human-nature relations in Pala. Fluid interactions and intimate co-habitations between humans and other species are the norm, wherein even the concept of ownership is undermined by such statements as, 'The mynahs [a local bird species] are like the electric light ... they don't belong to anyone' (Huxley 2009, p. 15). The island's human denizens are taught from early childhood that 'snakes are your brothers; snakes have a right to your compassion and your respect' (Huxley 2009, p. 194). In this vein, the following exchange is telling:

> A moment later a large green parrot, with white cheeks and a bill of polished jet, came swooping down from nowhere and landed with a squawk and a noisy fluttering of wings on Vijaya's shoulder ...
>
> Will: 'You people seem to be on remarkably good terms with the local fauna.'
>
> Vijaya: 'Pala is probably the only country in which an animal theologian would have no reason for believing in devils. For animals everywhere else, Satan, quite obviously, is Homo sapiens.' (p. 186)

Thus, it appears as though other terrestrials in Pala, even snakes which are typically feared and hated, enjoy considerable freedom of movement, and aren't systematically 'managed' or otherwise exploited. Perhaps this is due to the central role that ecology plays in Pala's education, wherein it is taught as a science as well as an ethic:

> Never give children a chance of imagining that anything exists in isolation. Make it plain from the very first that all living is relationship. Show them relationships in the woods, in the fields, in the ponds and streams, in the village and the country around it ... And ... we always teach the science of relationship in conjunction with the ethics of relationship. Balance, give and take, no excess ... (p. 211)

Curbing excess is a guiding principle that applies to dealings with nature and other species, as it does to Pala's socioeconomic and political organisation, as noted above. Though Pala as a utopian 'oasis of freedom and happiness' (p. 58) doesn't remain so for long; already early in the novel it is described as under threat internally and externally, by the influence of mass production and consumption, the 'wave of the future' in the form of crude petroleum, and dystopian forces steadily 'closing in' on the utopian enclave (pp. 58, 59). Yet, as Huxley observes, such an experiment is not about finding the best of all possible worlds but, through collective efforts and ceaseless trial and error, creating pockets of 'the better' that, however fleeting, remind us of the enduring possibility of other ways of being.

Ecotopia

Many ecotopian texts rightly designate, not a blanket 'humanity', but more specifically 'industrial human activity' – or the 'juggernaut of economic growth' at the heart of industrial-capitalist societies (historically situated in the Global North) – as a key driver of contemporary socio-ecological disintegration. This core critique is no less central in *Ecotopia* (1975), a 'modern classic' in the green utopian literary tradition (de Geus 1999) written by former editor and film critic with the University of California Press Ernest Callenbach. The novel is set a few decades into the near future after Northern California, Oregon, and Washington have jointly seceded from the United States to create an egalitarian and 'steady-state' ecotopian society, spearheaded by a 'Green' ('Survivalist') party who've implemented various transformations for socio-ecological sustainability.[2] Will Weston, the novel's protagonist and the first American allowed into the newly formed country, fulfils the important (utopian) critical role of the 'visitor'. In a telling early exchange, Ben, one of Ecotopia's denizens, remarks:

> ... Americans and their technology had been in the forefront of this tragic and irreversible process [biodiversity decline]. And indeed I hadn't realized how far it has gone: it is a horrible story. Our role in it was heavy, and thousands of marvellous creatures that once inhabited this earth have now vanished from the universe forever. *We have gobbled them up in our relentless increase* [emphasis added]. There are now 40 times more weight of humans on the earth than of all the wild mammals together! (p. 76)

---

2   Callenbach's prequel *Ecotopia Emerging* (1975) details the various struggles of the rebellious Bolinas who 'learn to say No' and whose successful defiance of the status quo ultimately leads to the establishment of Ecotopia (Buhle 2001, p. 152).

In this excerpt, Callenbach is alluding to unchecked socioeconomic growth as well as techno-capitalist industrial expansion. However, more problematically the end of the excerpt seems to refer to unchecked human population expansion. Whereas, as denoted in previous chapters, the ways in which we live, produce and consume – i.e. within capitalist systems predicated on endless expansion, and which encourage extreme concentration of wealth and resources in the hands of a small minority – are far stronger predictors of environmental despoliation than mere numbers. Worse than this, the 'Malthusian spectre of overpopulation' is a 'profoundly depoliticizing approach' (Ojeda et al. 2020, p. 316) that reinforces entrenched inequalities by perpetuating the scapegoating of racialised and gendered others as culpable for contemporary environmental ills, and therefore as acceptable sacrifices on the path to ecological salvation. The Lockean proviso (1690) that rights to the acquisition of land, resources, and wealth hold only to the extent that 'enough and as good' is left in common for others has been grotesquely violated, with extreme acquisitions by the few increasingly predicated on theft of the future itself from existing and future generations. As discussed throughout, socioeconomic inequality is not only ethically but also ecologically problematic, as it exacerbates ecological degradation through, among other things, placing pressure on the 'have-nots' to enhance their social position through increased material consumption (Wisman 2011).

Indeed, this dynamic isn't lost on Callenbach. In *Ecotopia* corporations and all productive enterprises, composed non-hierarchically by equal partners rather than high-earning executives and underpaid employees, are taxed according to their respective levels of earnings, and the government more generally implements long-range economic policies which call for 'diversification and decentralization of production in each city and region' (p. 8). As noted, *Ecotopia* is organised along a 'stable-state' basis, wherein plastics, for instance, are derived from living biological sources such as plants, and the production of non-biodegradable and/or resource-intensive goods – along with energy and material throughput more generally – is considerably restricted (Callenbach 1975, p. 83). All food waste, sewage, and garbage are 'turned into fertilizer and applied to the land, where it would again enter into the food production cycle', and recycling is compulsory (Callenbach

1975, pp. 18–20). In Ecotopia, humans' position more generally within the wider mosaic of terrestrial life forms is described as follows:

> Humans were meant to take their modest place in a seamless, stable-state web of living organisms, disturbing that web as little as possible. This would mean sacrifice of present consumption, but would ensure future survival ... People were to be happy not to the extent they dominated their fellow creatures on the earth, but to the extent they lived in balance with them. (1975, pp. 47, 48)

The idea is thus to scale back the 'human' enterprise, and to reduce harm and destruction wherever possible so that conditions of liveability for other present and future generations might be maintained. However, other terrestrials don't feature as central characters in the novel, and few details are offered by way of their lived experiences in this Ecotopia; although, as denoted above, a positive qualitative shift in how humans relate to other terrestrials and the wider biosphere is apparent and portrayed as a central feature of the good life. Emphases on cultivating relations with other species that are less exploitative and dominating persist throughout, as well as the need to decentre the 'human' by underscoring the myriad ways in which humans are thoroughly dependent upon – and embedded within – wider multispecies assemblages. For instance, other terrestrials are deemed 'fellows', and trees are often referred to as 'brother tree', suggesting the presence of more egalitarian multispecies relations. Clear-cutting forests has been banned, and wood harvesting is offset by work in a 'forest camp' to replace wood that is consumed (p. 55). Cows, no longer confined within the hellish abysses of factory farms, roam freely throughout the countryside. Moreover, the high-speed electric trains in *Ecotopia* are filled with plant species of various sorts (p. 8), rivers flow freely through major cities, and trees are abundant.

Ecotopia's influence wasn't limited to the burgeoning environmental movement in the West. It also made waves in the mainstream. As Lohmann (2018) observes, fourteen years after Ecotopia was published, NBC News anchor Connie Chung referred to the novel's visions as 'alarmingly prophetic' (p. 179; Oliver 1989). This was amidst a wider context increasingly beset by pollution, climate change and the threat of nuclear war. Lest it appear that ecotopian texts exist abstractedly outside of 'real world' social and political affairs, Lohmann (2018) further points out that many seemingly fictional ecotopian values and practices have been 'adopted' into mainstream cultures, including recycling, free public transport, biodegradable plastics and universal basic income schemes, to name but a few (p. 180; Samuel 2020). Others (Timberg 2008) have cited *Ecotopia* as a key inspiration behind bioregionalism, a movement that espouses an 'ecotopian worldview' marked by place-based sensibilities and a 're-grounding of culture and community within particular watersheds and ecosystems' (Lockyer & Veteto 2013, p. 8).

Pacific Edge

Kim Stanley Robinson is a major figure within the fields of green utopian and speculative fiction. His notable *Mars* trilogy features detailed explorations of the politics and logistics of world-building beyond terra, and prescient musings on the ethics of terraforming. Like H. G. Wells, a palpable utopian current pervades many of his works. In Blochean fashion Robinson's utopias, from the local to the cosmic, variedly 'grow out of the material conditions and contradictions of [his] alternative worlds' (Moylan 2021, p. 121). Robinson's 'transformative horizon of radical democracy, socialist/mixed economics, liberated culture and permaculture' (Moylan 2021, p. 144) features prominently in his ecotopian novel *Pacific Edge* (1990), the third instalment in his *Three Californias Trilogy*. The novel depicts a utopian society called El Modena set in Orange County, California in 2065, wherein tensions between growth-oriented and steady-state socioeconomic systems in tune with ecological integrity emerge as a central problematic. *Pacific Edge* especially evinces what Moylan (2021) aptly denotes as Robinson's 'ecologically informed utopian thinking' which variedly develops holistic critiques of the present world whilst sketching radical visions for a transformed socio-ecological reality in which humans are 'part of nature and, though important agents, not dominant' (pp. 138, 139). The story of El Modena is told largely through the eyes of its main protagonist, Kevin, whilst his grandfather

Tom facilitates cognitive estrangement by ruminating on the stark contrast between the ecotopian society of El Modena and the previous capitalist order:

> I hated capitalism because it was a lie! It said that everyone exercising their self-interest would make a decent community! Such a lie! It was government as protection agency, a belief system for the rich. (p. 56)

Like other ecotopian texts and social movements (Chapter 5), *Pacific Edge* designates global capitalism and its catastrophic ruination of earth's dynamic assemblages as a dystopian force par excellence. Although El Modena is no post-capitalist society, it is organised along decidedly more egalitarian socioeconomic lines, wherein citizens are taxed more heavily as they approach the system's personal income cap, and funds amassed therefrom are used in order to fund local social services. Individuals or corporations that attempt to syphon their wealth to off-shore tax havens have their assets seized and redistributed within local communities (p. 91). Across genres there is an awareness of the destabilising and unethical socio-ecological effects of extreme inequities in wealth and access to resources. Robinson also describes daily life in El Modena in exquisite detail, from the particular sights and scents of the region's flora and fauna, to the intricacies of the messy and sometimes mundane political deliberations that make the collective.

In terms of intraspecies relations in *Pacific* Edge, the agency and wellbeing of other species don't feature explicitly. However, a 'land ethic' (Leopold 1970; Moylan 2021) wherein humans are depicted as members *of* and not apart from the wider biotic community runs palpably throughout. Buildings and dwellings in El Modena follow biophilic design principles, wherein plant communities are incorporated into human homes and all-encompassing glass walls blur distinctions between the buildings' 'inside' and 'outside'. Moreover, key political deliberations in El Modena take place via local councils, wherein specific groups are tasked with representing the interests of other terrestrials. For instance, during one meeting after someone proposes to cut down trees bordering a local reservoir, a member of the Wilderness Party counters the motion, arguing that razing the sacred trees would constitute a 'wanton act of destruction' (1990, p. 26).

Moreover, inklings of the emergence of an 'ecological self' (Naess 1995) can be detected in the following observation by Kevin:

> He knew the configuration of every dark tree he passed, every turn in the path, and for a long moment rushing along he felt spread out in it all, interpenetrated, the smell of the plants part of him, his body a piece of the hills ... (p. 32)

In this passage as well as in other similar instances throughout the novel, Kevin articulates an ecocentric identity or 'ecological self', a term first coined by eco-philosopher Arne Naess (1995) to denote an expansive sense of self that extends beyond the individual 'self' construct in its identification with wider ecological assemblages. The 'ecological self' and associated processes of cognitive and perceptual identification in turn yield a deep sense of empathic connectedness and emotional resonance with others, particularly when they are perceived to be under threat. This is perhaps why Kevin becomes a key figure in the battle to protect Rattlesnake Hill – one of the last areas of intact wilderness in El Modena – from encroaching pro-development forces within the community. Although, there is the risk with this mode of identification that the 'other' merely gets subsumed within the expanded 'Self', rather than acknowledged and cherished as a singular entity. Nevertheless, with care, the 'ecological self' holds promise as a source of environmentally responsible behaviour, and as a way to help promote empathic engagement with the lived experiences of other beings.

Robinson's *Pacific Edge* depicts no Edenic ecotopia. As Lucy Sargisson (2012) aptly observes, this ecotopia is 'self-consciously flawed, containing malcontents and transgressors of its own internal value system' in order to 'caution against complacency, especially about the belief that we have arrived at the utopian destination, understood nature and can therefore represent it' (p. 130). And that's just as well, for there is no perfection to be had in our dealings with others. This work reminds the reader that the complex politics of cohabitation and ecotopian world-building is never finished or straightforward, often involving the negotiation of conflicting interests and perspectives that are difficult to ascertain at the best of times.

Always Coming Home

Ursula Le Guin's incredible breadth of utopian and science fiction works are pervaded by key themes such as decoloniality and, crucially, suffused by indigenous knowledge and ethics informing humans' intimate entanglements with agentic terrestrial kin. The latter can in part likely be attributed to Le Guin's upbringing, both of whose parents were anthropologists. Traces of these origins can also be seen in the narrative form of many of her works, which often feature richly detailed descriptions of people, places, and relationships in line with the classic style of ethnographic writing (Baker-Cristales 2012). Le Guin's singular ecotopian work *Always Coming Home* (1985) has been described as a 'postapocalyptic coming-of-age tale' set in a future, post-industrial Northern California centred around an ecotopian community evincing Native American and other indigenous cultural practices (Baker-Cristales 2012). Much of the story is narrated through the lens of the novel's protagonist, a girl named Stone Telling, whose mother is from the (ecotopian) Kesh community whilst her father is of the (dystopian) patriarchal, avaricious and war-like Condor people. The novel's fragmented and non-linear narrative structure is often punctuated with intricate footnotes and images explaining Kesh words, ideas, kinship practices, and the region's local flora and fauna. At other times the story is told by an anonymous ethnographer figure known as 'Pandora'. The novel is populated by all manner of people and perspectives who contribute to the story's rich narrative arc.

Always Coming Home features decidedly non-hierarchical human interactions with other species, and in my view offers among the best examples of what Moylan (2021) terms a 'more inclusive ontology of becoming utopian' (p. 151). The Kesh, who inhabit an area known as the Valley, exhibit a vital-materialist cosmological system wherein even stones and other 'non-living' components of the natural world are brimming with agency. Whilst on an expedition through a forest as part of a rite of passage into adulthood, Stone Telling remarks: 'It was a long way I went before a spring *let me find it* [emphasis added]' (p. 20). Here, the notion of the human subject as the only active agent navigating a world of mute objects is dismantled, with the human re-situated in a world full of other persons who must be negotiated and reckoned with. The Kesh's seemingly anti-hierarchical and non-dualistic ontology encompasses the 'Nine Houses of Being':

> The beings or creatures that are said to live in the Five Houses of Earth and are called Earth People include the earth itself, rocks and dirt and geological formations, the moon, all springs, streams, and lakes of fresh water, all human beings currently alive, game animals, domestic animals, individual animals, domestic and ground-dwelling birds, and all plants that are gathered, planted, or used by human beings ... The people of the Sky, called Four-House People, Sky People, Rainbow People, include the sun and stars, the oceans, wild animals not hunted as game, all animals, plants, and persons considered as the species rather than as an individual, human beings considered as a tribe, people, or species, all people and beings in dreams, visions, and stories, most kinds of birds, the dead, and the unborn. (p. 44)

Kesh expansive cosmology and kinship networks thus include not only all living and non-living entities as beings on equal ontological footing 'dancing the same dance' but also non-corporeal entities such as 'beings in stories or dreams' (p. 47). As discussed in the preceding chapter, all 'actants' have a capacity to affect and be affected by others, wherein none enjoy special ontological status in terms of exclusive possession of the ability to engage in meaningful actions. Reality is thus (re)conceived as a vast, pluralistic and dynamic network of ever-shifting alliances between various actants who gain as well as lose power through such alliances, indeed are their alliances, and wherein humans are thoroughly dependent on other actants for their existence:

> Thinking human people and other animals, the plants, the rocks and stars, all the beings that think or are thought, that are seen or see, that hold or are held, all of us are beings of the Nine Houses of Being, dancing the same dance. (p. 307)

Despite the quote's ending, traces of the Cartesian subject–object divide re-emerge when certain beings are posited as either seen or seeing, thinking or thought. Presumably at least, most animals can be both seen and seeing, thinking and thought, and not merely one or the other. Nevertheless, the novel contains trenchant criticisms of the possessive ('having') orientations (Fromm 2013) characteristic of industrial capitalism, that is, the desire to acquire, consume and dominate. The Kesh's strong aversion to the concept of private ownership, and the vibrant animism that infuses their language, have led them to substitute the word 'pet', which has 'patronizing and condescending overtones', with the word commensal, which denotes 'people living together' (p. 419). Thus, traditional hierarchies between humans and all other beings are critiqued and dismantled to a considerable degree, whilst human modes of relationality with other species are informed and characterised by egalitarianism, reverence and respect.

Though constituting notable strides towards non-exploitative and non-oppressive multispecies relations, animal agriculture, hunting and the use of other animals to meet human needs and desires persist in *Always Coming Home* as well as in the other novels in question. Animals such as chickens and fish that have been traditionally categorised as 'food' remain as such. In *Always Coming Home*, deer and quail are hunted and domestic animals such as sheep are featured in festivals, while 'sheepskin and wool [are] used for various leather goods and such' (p. 414). However, though sheep are utilised for various human purposes, they are 'not a symbol of passive stupidity and blind obedience, but rather ... regarded with a kind of affectionate awe, as an intrinsically mysterious being' (p. 415). Thus, other species are treated as both respected commensals as well as sources of warmth and nourishment for human sustenance. This alludes to a contentious issue within environmental-political discourses that warrants further discussion: are ecotopian relations with fellow terrestrials compatible with our consumption and use of them for food and the like, or is such use inherently exploitative and therefore ethically unacceptable?

Interlude: Plumwood's Theory of Ecological Animalism

Val Plumwood's contextual theory of ecological animalism (2003) seeks to dismantle human supremacy by simultaneously resituating humans in ecological terms and other species in cultural and ethical terms. This is done by (re)affirming a universe wherein other species as well as humans are equally available for mutual and respectful use. Plumwood refers to the unyielding opposition to consumption or use of any kind as a moral wrong in and of itself – rather than by virtue of its mode and extent – as ontological veganism. An ontological vegan might argue that it is inherently wrong for humans to consume other animals, that the very nature of their being should exempt them in absolute terms from consumption. However, for Plumwood such a position works to reinforce and reproduce human/animal and nature/culture dualisms. By framing only beings within arbitrarily delineated bounds of ethical consideration as inedible (i.e. animals), boundaries which by their nature exclude (i.e. plants), ontological veganism denies (humans') ecological embeddedness. Illusory rigid boundaries between sentience/nonsentience, animals/plants are reproduced in order to justify claims for exempting and extricating certain beings from wider food webs, a move that is ontologically problematic because the dynamic assemblages and intra-actions that constitute terrestrial existence don't conform to such rigid categorisations. Moreover, ontological veganism risks demonising predation more generally by positing consumption as inherently unethical, a position that can only be maintained by redeploying yet another binary: 'Nature/Culture'. Herein, humans as the sole 'cultural' beings are deemed capable of wholly abstaining from consumption of others, while other-than-human predators are consigned to the 'Natural' realm of necessity. Yet, Terra is a composite of a vast multiplicity of complex and ever-shifting relationships, the consumption of others being a particularly crucial one: there is no shortage of evidence demonstrating the catastrophic consequences in the form of trophic cascades and other far-reaching ramifications often following the absence of predators like Orcas and Wolves from the ecosystems within which they're embedded (Borrvall & Ebenman 2006).

Use and consumption of others are integral, inescapable aspects of ecosystem functioning, and are not inherently exploitative or reductive. All animals (even some plants!) must take life in order to live. When we die, we become food for a host of microorganisms, nourishing them in turn. Similarly, consider mutualistic forms of 'use' between plants and pollinating insects, or among sea anemones, anemonefishes and zooxanthellae. Conversely, reductive or exploitative forms of use involving attempts to exercise domination over others, and use them as mere means to one's ends, are the kinds that Plumwood and many of the texts discussed problematise and seek to dismantle. Keeping our terrestrial kin confined for most of their lives so that we can harvest their skin, fur, flesh and other parts for our own consumption or entertainment, wherein many can't interact with their conspecifics in ways that they'd wish or even see the sun, constitute the exact opposite of mutuality and respect. They are thoroughly degrading and profoundly inimical to the formation of ethical interactions with our co-evolutionary kin, hence Vijaya's apt reference to Homo sapiens as 'Satan' as far as most other species are concerned. The industrial-agricultural complex has poisoned waterways, razed forest homes, so altered the climate as to create new generations of other-than-human climate migrants, and systematically exterminated them to the extent that our biomass and that of our enslaved, factory-farmed kin now exceeds that of all other vertebrates apart from fish. Such systems have effectively stolen other terrestrials' abilities to flourish in the present as well as into futurity. Alluding to the theme of ownership, often a figure of criticism in these ecotopian works, Plumwood problematises instrumental and reductive forms of use and modes of relationality characterised by Fromm's (1967) 'having' orientation. Here, the overarching objective is to acquire and possess, a key impediment to the construction of more ethical modes of relationality with our co-terrestrials. Indeed, for Tree, one of the REA participants discussed in the following chapter, it's not the consumption and use of other species that's ethically unacceptable but rather presently institutionalised modes of domination and exploitation in the form of industrial factory farming:

> I think there are some uses of animals I could support and there are some I wouldn't support, and there are some I've yet to make my mind up. I mean, using animals for sport and for circuses I think is completely unacceptable ... I think in terms of our use of animals, just to elaborate on that, most, well, not most because there are animals that just eat plants, but many eat other animals, there are lots of those, and we're essentially one of them. But what distinguishes us from others is that other animals eat what they need.

In other words, no living heterotroph can fully abstain from consuming and 'using' others; this is the very business of life. However, on the basis of the immeasurable cruelty involved in industrial factory farming, on its violent reduction of living beings to mere sellable commodities, and on its disastrous ecological impacts, calls for abstaining from or at least radically reducing meat and dairy consumption find firmer ground. In this vein, veganism ought to serve as a powerful mode of political resistance against systemic cruelty and harm, not an attempt at purity which corrodes effective political action whilst rehashing the very dynamics of violent exclusion that it purportedly seeks to challenge (Shotwell 2016). The following excerpt from *Always Coming Home* poignantly illustrates the essence of Plumwood's theory of ecological animalism:

> Come among the unsown grasses bearing richly, the oaks heavy with acorns, the sweet roots in unplowed earth. Come among the deer on the hill, the fish in the river, the quail in the meadows. You can take them, you can eat them, Like you they are food. They are *with you, not for you* [emphasis added]. Who are their owners? This is the puma's range, this hill is the vixen's this is the owl's tree, this is the mouse's run, this is the minnow's pool: it is all one place. Come take your place. No fences here, but sanctions. No war here, but dying; there is dying here. Come hunt, it is yourself you hunt. Come gather yourself from the grass, the branch, the earth. Walk here, sleep well, on ground that is not yours, but is yourself. (pp. 76–77)

The notion that the ground 'is not yours, but is yourself' recalls the disavowal of ownership of earth others in *Island*, as well as Kevin's intimation of an 'ecological self' when he sensed his body as 'a piece of the hills'. It is also a common understanding of land seen in many indigenous cultures (see Chapter 6) wherein one is not apart from or even on but inextricably *of* the land. Other terrestrials are, crucially, with us, not for us; if humans are indeed animals who differ from other species only by degrees rather than kind, as actants on equal ontological footing, then surely, like them, we too are food – as well as so much more. Other terrestrials are our commensals, each deserving of respect on the basis of their fellowship with us. None of us may be extricated from biospheric interconnectivity or elevated to a position of superiority. Living well and ethically with earth others, particularly amid the turbulent Capitalocene, demands a recognition of and respect for their 'otherness' which is not reducible to our desires, or ours to do with as we please; that they are more than mere means, but whole worlds in themselves.

The Fifth Sacred Thing

> We hope for a harvest, we pray for rain, but nothing is certain. We say that the harvest will only be abundant if the crops are shared, that the rains will not come unless water is conserved and shared and respected. We believe we can continue to live and thrive only if we care for one another. Este es el tiempo de la segadora [this is the age of the Reaper], when we inherit five thousand years of postponed results, the fruits of our callousness toward the earth and toward other human beings
>
> – Starhawk (1993, p. 17)

A segadora or 'Reaper' is an agricultural instrument used for reaping ripe harvests. It is also, perhaps more notably, a reference to the personification of death seen in many cultures – i.e. the Grim Reaper who arrives to collect human souls after death. That which is sown is ultimately reaped; if the sower or sowers are long gone, others must inevitably inherit the harvest.[3] Such is the historical backdrop informing *The Fifth Sacred Thing* (1993), an ecotopian novel by American activist and eco-feminist scholar Starhawk. The novel depicts an ecotopian enclave referred to as 'the City' set in San Fransisco, California[4] in 2048, wherein its inhabitants must continually contend with – and collectively reap – the consequences of ecological neglect sown by previous generations. The story is told largely through the perspective of a woman in her 90s named Maya and some of her surviving kin, including Madrone and Maya's grandson Bird. The

---

3   In Philip Dick's (author of the famed 1968 sci-fi classic *Do Androids Dream of Electric Sheep?* that inspired the film Blade Runner) eco-science fiction short story 'Survey Team', a group of seemingly human explorers fleeing a dead terra in search of a new habitable planet reach Mars, only to find that it a graveyard, a dead world used up and 'sucked dry' (Dick in Ashley 2020, p. 30) by a long-since departed Martian civilisation. Towards the end of the story, as the explorers are searching for the Martians' escape planet, they realise that it was terra, and that they are in fact the descendants of the Martian civilisation 'back to reap the crop our ancestors sowed' (Dick in Ashley 2020, p. 36).

4   Callenbach's, Le Guin's and Robinson's ecotopias are also, curiously, set in California. Apart from the fact that they're all American, perhaps it is California's mild climate and picturesque geography, with towering red woods, that continues to captivate and seem an ideal place to build an ecotopian enclave.

City boasts abundant streams, gardens and lush greenery, a fragile prosperity that is carefully maintained by the art of 'unwaste' (p. 3), wherein the inhabitants have learned how to rebuild a precarious existence on a land ravaged by fires and droughts. The essential foundations of life – the water, air, earth – are thus deemed 'sacred', and that which is sacred 'can't be bought or sold. It is beyond price, and nothing that might harm it is worth doing' (p. 18). This is the central theme that runs throughout the novel, as well as the guiding ethical principle that underpins intraspecies relations and the City's wider socio-political organisation more generally:

> The Earth is a living, conscious being. In company with cultures of many different times and places, we name these things as sacred: air, fire, water, and earth … to call these thing sacred is to say that they have a value beyond their usefulness to human ends … no one has a right to appropriate them or profit from them at the expense of others. Any government that fails to protect them forfeits its legitimacy. All people, all living things, are part of the earth life, and so are sacred. No one of us stands higher or lower than any other. Only justice can assure balance: only ecological balance can sustain freedom. Only in freedom can that fifth sacred thing we call spirit flourish in its full diversity … To honor the sacred is to make love possible. (Starhawk [Declaration of the four sacred things] 1993)

The City's socioeconomic and political organisation, predicated on mutual cooperation and sharing, revolves around the careful use of the four sacred things. As no one can assume private ownership over them, everyone has enough. Moreover, each is guaranteed a share of 'past wealth' and of existing resources, which translates into a basic stipend of credits for each citizen (p. 274). On Madrone's travels to Southern California to help rebel bands organising a resistance against the dystopian Millenialists[5] and Stewards, a Christian extremist movement that took over the government and abhors the Earth and all fleshly phenomena as synonymous with the Satanic (p. 29), the people she encounters are often incredulous upon learning of conditions in the City. In these regions ravaged by drought and deprivation, some have never even seen a real live tree (p. 299), and struggle to comprehend that water is free and doesn't have to be stolen.

---

5   The Millenialists' sacred principles, in stark contrast, include moral purity, racial purity, and spiritual purity (p. 272).

However, Madrone eschews the label of a 'perfect communist society' as a description of the City:

> Old giant state socialist countries like the Soviet Union failed because they attempted too much control over complexity ... [which] traces back to the mechanistic philosophy of the Enlightenment, which saw nature as a great machine, something we could ultimately know and completely control. (p. 273)

In contrast, 'nature' and the Four Sacred Things in the City are agentic and dynamic, ultimately beyond human attempts at manipulation and domination. The fundamental sacredness of the earth and inhabitants is also what informs their mode of political organisation and participation. At the Council, wherein the City's human denizens arrange to deliberate matters concerning the community and reach agreements based on consensus, there are some referred to as the Voices whose role is to speak on behalf of the Four Sacred things. For instance, there is 'the bearer of the Hawk mask ... guardian of the creatures of the air' and 'Salmon, guardian of the waters' who reminds humans of their responsibility to 'clean up the mess they made' (p. 50). Similarly, the Water Council bears the responsibility of purifying the city's toxic waterways so that the great salmon run might be restored (p. 45). Crucially, any decisions made at Council meetings must take into account other animals, plants, the waters, etc. (p. 46).

Regarding the use and consumption of other species, Madrone admits that the latter is the source of one of the longest-standing debates in the City:

> ... a lot of people think we all should be strict vegetarians ... but a lot of us believe we can't do without animals – not just for meat but for the part they play in the whole system. We need their waste for fertiliser; we use every part of the ones we kill ... We all believe the animals have to be treated well when they're alive and killed with the least possible suffering, with rituals to honor their spirits. (p. 186)

As alluded to above in the discussion of Le Guin's *Always Coming Home*, herein lies a system of total integration, with no strict hyper-separation between 'humans' and other species or 'humans' and 'nature'. Even if it were desirable, no entity could be removed from the wider nexus of relations within which they're enmeshed. As such, embedding ethical principles of care and reduction of harm in their dealings with one another helps facilitate more respectful intraspecies relations. An important additional step in this direction would involve not merely enquiring about what we need them for, but also what they might need from us in order to flourish. A related motif throughout the novel is that of groundedness, of the dangers that stem from being too far removed, both physically and psychologically, from terra. After a decade-long imprisonment by the Stewards, Bird revels in the opportunity to once again go 'outside' and:

> stand once again on the living earth, to feel her like a vibrating body under his feet, to breathe air and feel a wind on his face ... to smell the compounded incense of leaves, dust and ocean salt ... to see living things in their soft colours, blue and green, and umber earth below. (p. 58)

Bird and his family are Pagan witches and dreamers who occupy, and are especially attuned to, the porous boundary between the world of physical forms and 'the realms of energies and spirits that underlie the physical' (p. 171). They frequently commune with the spirits of departed kin, asking for guidance in relation to daily affairs, similarly to the expansive kinship networks of the Kesh in *Always Coming Home* which include past, living and future relatives. By living in right relation to the Four Sacred Things – i.e. honouring, respecting, protecting and sharing them – one can gain access to the Fifth, spirit, which together form the circle of life (p. 300). However, towards the end of the novel it's revealed that, more precisely, love is the fifth sacred thing (p. 473). Love is the life-affirming force that can shatter the confines of the most violent and oppressive realities, and guide our steps towards better worlds.

This dynamic ecotopia is thoroughly *in* and *of* terra, what I refer to as a terrestrial modality of utopianism (see Chapters 3 and 8) that is rooted firmly in (respectful) relation and entanglement with our terrestrial kin. Like the REAs discussed in the following chapter, it does not seek to beam its vision of the good life into a distant 'beyond', whether spatially or temporally, but grounds itself in the now 'here below' (Latour 2017). Yet, this ecotopian enclave lives amidst constant potential peril from its dystopian surroundings – from unknown viruses, radiation exposure from ozone depletion, the effects of the wider ecological damage still being wrought without, and anti-utopian forces from the South – always threatening to close in. These forces shatter any illusions of stasis or total isolation. There are no rigid boundaries, whether between humans and other terrestrials or between the ecotopian society and the wider, deficient reality within which it has carved a tentative oasis of radical possibility. Starhawk's novel viscerally highlights the fears, tensions and ongoing challenges associated with 'building the better', a process that is never finished. Yet wherever such spaces emerge, they 'demonstrate hope' that 'people can live well together, can care for one another, can heal and build in harmony with what is around us' (p. 238). If nothing else, they deal yet another major blow against the 'war of imagination' waged by the deficient present which, by limiting dreams and visions, makes us 'accept within ourselves its terms, to believe that our only choices are those that it lays before us' (p. 238).

(Posthuman) Urban Futurities

> How to conjure up a picture, for instance, of a town without pigeons, without any trees or gardens, where you never hear the beat of wings or the rustle of leaves – a thoroughly negative place in short?
>
> – Camus (1948, p. 5)

Though now home to the majority of the world's human population, cities – like life itself – have always been multispecies endeavours. In Nottingham, where I have lived for seven years, I am honoured to share this urban landscape with peregrine falcons, pigeons, Canada geese, swans, foxes, crows, magpies, rats, mice, orb-weaver spiders and many other creatures – not to mention the incredible array of plant life, from birch trees to sugar maples that vibrantly herald the arrival of autumn each year. What an incredible, humbling array of perspectives on the world (Hunt p. 206) they provide. The above excerpt is Albert Camus's description of Oran, the site of a devastating plague outbreak in his seminal work, *The Plague* (1948), which portrays an urban landscape that can be described as decidedly dystopian due to its marked absence of other-than-human life. And this is precisely what the visionary multidisciplinary artistic movement Solarpunk seek to counter in their (ecotopian) publication *Multispecies Cities: Solarpunk Urban Futures* (2021), a timely collection of imaginative essays and short stories that seek to (re)envision cities as living spaces of multispecies flourishing. The entries in the collection variedly explore the joys – as well as trials – of living with the myriad terrestrial kin we share our planet, and increasingly our urban dwellings, with. *Multispecies Cities* is an important and timely work in its explicit attempts to counter dominant (and dystopian) technocratic, capitalist and humanist visions of cities as spaces crafted for – and populated almost solely by – humans (Rupprecht et al. 2021, p. 4).

Rather, from the perspective of characters situated in diverse spatial-temporalities, this collection presents a mosaic of visions of more inclusive and diverse urban spaces populated by all manner of purposive entities who speak – if only humans make the effort to listen well and deeply. One needn't understand the language of others in order to follow them or get a sense of what their needs and desires might be. What is needed, as the collection variedly suggests, is attention and imagination (p. 211). Attentiveness, remarks Despret, involves '"giving your attention" to other beings and at the same time acknowledging the way other beings are themselves attentive' (2022, p. 5). It involves the acknowledgement of other beings as worlds in themselves with their own unique and valid ways of knowing and being. As discussed in Chapter 2, plants, for instance, are far from mute or lacking in agency but 'speak through *space*, in the manner they open their leaves, spread petals, turn to the light [known as "phototropism"] …' (Chabria, p. 16). When they are in want of hydration, light or other necessities, they tend to wilt and lose their vibrant demeanour; by learning how to recognise such tell-tale signs, we can be better relatives and help them thrive.

The collection's recurrent theme of interspecies communication is especially prominent in Andrew Hudson's poignant exploration of human relations with mammoths spliced from elephant genes in 'The Mammoth Steps'. Herein, the elephant-mammoth hybrid Roomba, through careful observation and prolonged, intimate contact …

knew Kaskil's body-language, recognised many words and gestures, and likewise could signal his feelings and opinions with a nod or trunk swing, a trumpet or harrumph, or the thrumming infrasonic rumbles that Kaskil's phone registered as pictographs or emojis. (p. 236)

Attentiveness and, in this case, a little technological assistance, can help facilitate more effective interspecies communication. However, limits to our abilities to fully understand singular others always remain. E. H. Niebler's story 'Crew', human protagonist Qiqi is able to establish some form of communication with Old Woo, an ex-military sperm whale, via technological means, but only just:

Old Woo made his depreciating harrumph. Qiqi had no doubt that the fullness of its meaning could not ever be adequately captured by the translation harness permanently fixed to the massive block-shape of his head. (p. 274)

Though the inner lives of others to an extent always remain a mystery, a partial mutual understanding is possible across such divides when one strives to listen and become attentive to others and their own ways of communicating. *Multispecies Cities* presents decidedly posthuman visions of urban spaces (Braidotti 2013), with drones described as 'beings' (Spires, p. 94) and robotic dogs on Mars with uncannily dog-like mannerisms (Stevens, p. 191). The collection above all urges us to embrace an ethic of care and responsibility for the more-than-human beings with whom we are always inextricably entangled, even those not considered 'alive' in the conventional sense.

In the story 'Vladivostok', two Metropolis[6] enthusiasts from Canada set off to the Russian wilderness to film the elusive Amur tiger so that it can be incorporated as a new species entry into the game. For Bryan, Metropolis is in many ways more alluring than real life, where 'It has everything you want. And we can build nature there, right?' To which his partner Ronan responds, 'But when we play Metropolis, we're actually there. The animals aren't. They're just programs' (Balwit, p. 71). This interchange is especially telling within the context of the sixth mass extinction and the mounting

[6] Metropolis is a city simulator game wherein participants can engage in the construction, management and politics of their desired urban spaces.

loss of our terrestrial kin. Attempts to digitally preserve threatened species, or resurrect those long gone through genetic engineering feats such as 'de-extinction splicing' (Hudson, p. 236), as in Hudson's 'The Mammoth Steps',[7] don't revoke the loss of these singular evolutionary achievements. Moreover, what does it mean for an elephant-mammoth hybrid to come into a brave new world so altered in relation to that which nourished their ancestors? (Hudson, p. 238). This notion is also explored in D. K. Mok's poignant story 'The Birdsong Fossil' wherein the protagonist Yuzuki, continually haunted by the seemingly unstoppable losses of species sweeping our world, reflects on what extinction truly entails and what it is that we actually 'bring back' when we resurrect an extinct species:

> You're giving people a false sense of security. 'If a species goes extinct, we'll just bring it back. No harm done.' And you know that's not true. Once we lose a species – it's gone. You might bring back the body, but you're not bringing back the mind, the culture, the ecological systems. You're not bringing back the dolphins who teach their daughters how to forage using sponge masks ... (p. 297)

Not even the most sophisticated virtual reality game or lavishly-funded modern genetics lab (Minteer 2015, p. 15) can recreate or substitute what's been lost – singular constellations of memories, experiences, capabilities, and personalities. What remains in their wake is an ineffable absent presence and present absence, a palpable 'gap in the world' (Lawrence in Muldoon 1998). You might be able to 'reanimate a corpse', but that doesn't 'bring back the *person*' [emphasis added] (p. 307) or the spatial-temporal relations that constituted them. What's more, what of the critically endangered species

---

7   Hudson's story might be closer to reality than initially anticipated. Recent breakthroughs in genetics and molecular biology have made it possible for scientists to use ancient DNA taken from museum specimens, sequence the extinct genomes and 'edit' the DNA of closely related species in order to produce a genetic blueprint very closely resembling the extinct forms (Minteer 2015, p. 12). The Texas-based biotech company Colossal, co-founded by genetics professor George Church, has devoted millions to bring back the thylacine and the woolly mammoth – in this case by creating mammoth – elephant hybrids through gene editing and the use of a surrogate elephant mother- as part of a wider effort to 'rewild' the arctic tundra (Morton 2022).

still with us? The resurrection of long-absent species does nothing to mitigate the root drivers of biological annihilation; in fact, it could potentially wreak havoc on existing terrestrials who've adapted to conditions now decidedly different from those that previous species thrived in (Minteer 2015). Rather, the millions of dollars currently being spent on Promethean de-extinction projects to 'recreate' species like the Woolly Mammoth and the Thylacine would be much better spent on radical measures to bring species back from the brink rather than the grave. Similarly, Yuzuki tries to preserve that which leaves no trace in fossil records (p. 303) – i.e. the 'courtship dance of the red-crowned crane', and similar stories that link species as intergenerational achievements across space and time (van Dooren 2014a) – by replicating animal cultures in robots.

Amin Chehelnabi's story 'Wandjina'[8] illuminates in heart-wrenching detail the incalculable suffering of the hundreds of millions of terrestrial victims of the wildfires that erupted across Australia in January 2020. The story's committed and occasionally misanthropic[9] protagonist Tara selflessly puts her own safety and wellbeing aside in order to rescue as many injured animals as possible. This is made all the more difficult due to the at times overwhelming grief she grapples with from seeing so many animal kin perish as a result of the self-serving and individualistic actions of the Australian government. The latter have designed cities and allocated scarce resources exclusively for 'Australians', which only includes the country's human inhabitants (Chehelnabi, p. 251). Using resources or equipment to help other terrestrials, on the other hand, carries hefty penalties and fines. When Tara and her other rebellious human comrades make it to an (illegal) animal rescue and recovery site, the scene described evinces a pocket of radical ecotopian potential of interspecies care amidst hellish surroundings (Kirksey et al. 2014):

---

8   Wandjina in Aboriginal mythology refers to cloud and rain spirits as powerful forces for change; in this story, the name is attributed to the carrier used in order to rescue the animal victims of Australia's climate-ravaged landscape.
9   This stems not only from Emma's first-hand witness of some humans' crimes against other animals, but also because of an internal conflict between the politics of her faith, Islam, and her unbridled love of other-than-human animals.

> *This place really is out of this world* ... so many veterinarians, well, they look like veterinarians! Several wallabies stood by, waiting patiently while several children prepared to feed them. A man and woman brushed a dingo down, removing ash and debris, and the dingo licked their faces in appreciation ... people ran around carrying equipment, and others held animals the same way Tara held that bandicoot ... (Chehelnabi, p. 258)

In a time and place wherein 'they're all dwindling' ... they're also 'hanging on. Fiercely' (p. 244) – hope clings tenaciously to life. Many other stories in the collection also highlight – though perhaps they might this might have been explored in even greater depth – the tensions and difficulties that attend to living with and caring for others. There is no perfection to be had in our dealings with other terrestrials; interspecies relations are always laden with triumphs as well as tribulations, joy and co-operation as well as conflict. In Vladivostok, violence between tigers and humans lessens, and a mutual respect is established, once humans are no longer forced out of necessity to encroach on the tigers' territories and hunt them (Balwit, p. 69). Interspecies care is also no smooth or effortless feat. As Chatzidakis et al. note in their poignant work *The Care Manifesto: The Politics of Interdependence* (2020), 'Attending to the needs and vulnerabilities of any living being ... can be both challenging and exhausting' (p. 23); moreover, care can also cause harm, as in the 'violent care' of conservation wherein predators and pests are culled in service of careful protection of other species (van Dooren 2014b). Like interspecies communication, care is complex, and requires continuous effort, attentiveness, as well as time and material resources. This is precisely the observation made by an elderly man meticulously tending to a rooftop garden in Singapore in the short story 'Untamed' by Timothy Yam: 'Taking care – not easy. I do for forty years already – every day in the sun, hard work to maintain this place, make sure the plants and animals are well' (p. 100). Or similarly in Octavia Cade's story 'The streams are paved with fish traps', care for an injured eel being temporarily housed in the human protagonists' home bathtub is described as 'a nuisance', forcing them to bathe out of their sink for the duration of the eel's recovery (p. 265).

The cosmopolitical approach outlined by Latour (2004b) and Isabelle Stengers (2010) seeks to deconstruct the profoundly humanist roots of traditional cosmopolitanism as espoused by Immanuel Kant (Watson 2014). Rather, the aim here is to reconceptualise the very notion of political life, namely by recasting the role of humans in multifarious political communities such that we see ourselves not as representatives, but as 'diplomats' (Latour 2004b). The role of the diplomat is, crucially, not to discover 'a common language, or an intersubjective understanding between protagonists constrained by diverging attachments and obligations' (Stengers 2010, p. 29), but to 'propose ways forward that are acceptable to the different parties, perhaps for very different reasons' (Celermajer et al. 2020, p. 13). This would require, among other things, concerted efforts to engage with radically different ways of knowing and being, while remaining critically aware of the limits of our capacity to fully understand the 'other' and of power dynamics implicated in acts of representation (Celermajer et al. 2020, p. 10). In a similar vein, John Dryzek (2007) calls for extending Habermas's central political notion of 'communicative rationality' beyond its reliance on (human) 'reasoned discourse' to include other forms of communication (i.e. non-symbolic) engaged in by other terrestrials so that they might more readily be included in political deliberations. Moreover, Andrew Dobson (2010) adds that a more inclusive and democratic terrestrial politics requires that we cultivate the art of listening (p. 766) – particularly to the needs of those residing beyond the bounds of a given collective. Crucially, this places onus on those who hold the power to be more attentive to excluded others. Similarly, Latour and Stengers pose such vital queries as, 'What are the diverse interests of all the human and non-human stakeholders? And perhaps most importantly, how do existing power dynamics constrain the possibility of some interests being represented or realised?' (Celermajer et al. 2020b). Thus, with respect to co-creating ecotopian collectives, we ought to continually keep at least two key queries in mind: (1) the epistemological one concerning how many actants there are to take into account, and (2) the ethical (or utopian) one of what the best of possible worlds *ought* to consist of (Latour 2004a, p. 93). In other words, the vital task is to continuously alert the ever-expanding collective as to who is left out:

> For them [moralists], the collective is always trembling because it has left outside all that it needed to take into account to define itself as a common world. A spider, a toad, a mite, a whale's sigh, these are perhaps what have made us fall short of full

and entire humanity, unless it was some unemployed person, some teenager on a street in Djakarta, or perhaps it was some black hole, forgotten by everyone, at the edge of the universe, or a newly discovered planetary system. (Latour 2004a, p. 158)

In Latourian cosmopolitical (Latour 2004b, p. 454; Stengers et al. 2010) fashion, Phoebe Wagner's 'Children of the asphalt' wonderfully illustrates the messy politics of cohabitation when a member of an unknown species (later named the 'Landrus') arrives in town and begins causing damage to local food sources. The human denizens, referring to all terrestrials as 'kin', have resolved never to kill or hurt any kin unless they display malicious intent. As such, they convene, taking as many voices as possible into account, to deliberate on *why* the Landrus might be flattening the town's crop fields, how much food and other resources this new kin member might need in order to thrive, and generally how the city's existing collective might make adjustments to their own habits so as to accommodate and respectfully live alongside this new kin member (pp. 164, 165).

Another prevalent theme is that of the transience of all things in this life, and how we ought to, following John Berger (2008), 'hold everything dear' whilst there is still time:

> I realise, they are terns. The scientist in me wants to jot down the observation, inform HQ and then do what ...? Write a paper about the return of terns? Petition to have the old species recorded? I marvel at the tern's beauty. This will suffice. Things are fleeting ... better to cherish them now. (Chng, p. 118)

Though one might add that marvelling at the singular beauty of the many creatures we share urban and other spaces with (Hunt, 206) will not suffice amidst these urgent times. Staying with the trouble (Haraway 2016) of the damage wrought by oppressive and exploitative systems such as capitalism and anthropocentrism on our terrestrial kin requires facing up to our responsibilities to them (Latour 2017). There is no going back to undo past harms, as many of the stories in the collection allude to. So, how do we make amends? How does one 'apologise to a river' (Victoria, 178)? How do those of us in 'shiny air-controlled cars and our plastered dome megawalls [who'd] been deaf to all songs ... listen to the songs of empathy again' (Taneja, 292)? By tuning in. By promising and continually striving to do better in the essential task of co-creating more liveable worlds in the here and now that take the interests, needs and desires of other terrestrials into more careful consideration. By embracing an ethic of care and responsibility (Chehelnabi, 246) for how our various doings in the business of living affect others, working to salvage those who still remain (Chenelnabi, p. 245), and recognising that they are worth fighting for. And by engaging in daily acts of kindness and compassion – small and large – which 'every day make the world a little better' (Mok, p. 321). In D. A. Xiaolin Spires' story 'The exuberant vitality of hatchling habitats', the two young protagonists create biodegradable structures to serve as new dwelling places for seagulls and other threatened sea birds; one such structure, named 'Promise Rock', signifies 'a pledge that we'll do our best' … to 'continue to treat this land, shared by humans, drones, gulls and other beings alike, with respect' (94). Forging more caring and respectful lifeways amidst the blasted landscapes (Tsing 2012) of the present might begin with such a promise, and the conviction espoused by *Multispecies Cities* that a better world is always possible.

What could ecotopian urban spaces that actively incorporate, nurture and respect other terrestrials look and feel like? Visions will vary, but fundamentally, there ought to be 'green spaces and trees in every street as we try to coax life back to the cities, to make it more than a two-dimensional space for a single species' (Cade, p. 270). As plant scientist Stefano Mancuso enjoins us in his imaginative work *The Nation of Plants: A Radical Manifesto for Humans* (2021), let us cover every available terrestrial surface with plants, a vision akin to that set out in Octavia Cade's entry 'The streams are paved with fish traps':

> … green walls and green roofs to, to encourage plant life and sustainable food sources, to draw the insects and the birds. The resting places built on skyscrapers, the road signs warning of safe nesting spaces for penguins. The ecological sanctuary within the city, where kiwi and pukeko are coming back. It's a harbour city, used to life in liminal spaces – seagrass and algae and shellfish in the intertidal zones, all of them encouraged. (p. 266)

Similarly, in Mok's 'The Birdsong Fossil':

> I could see the city's rooftop gardens blanketing the district like a mossy patchwork. From pocket community gardens fragrant with fresh herbs to sprawling manicured

parks frequented by the co-working cohort, these botanical eyries were interconnected by a tangle of wildlife bridges that facilitated non-human passage from roof to roof. (p. 298)

In Hudson's 'The Mammoth steps', elephants roam the streets of Phuket unimpeded and 'work with allied humans to construct elephant-sized buildings' (p. 239). *Multispecies Futures* begins with the fundamental premise that cities are dynamic assemblages of multifarious narrative agencies and lifeways (Steele et al. 2019), and proceeds to offer an array of visions of how such spaces could be (re)designed in order to help all of their inhabitants thrive. As homes to all manner of creatures, human and otherwise, some of whom occupy 'interstitial spaces between the facades of buildings (Steele et al. 2019, p. 412), as well as conceptually between 'here' and 'there', 'us' and 'them', cities evoke the possibilities of what Barbara Creed (2017) refers to as 'stray ethics'. Stray ethics highlights the shared cross-species experiences of estrangement, marginalisation and abandonment (p. 167; Steele et al. 2019, p. 412). The stray resides outside of the delineated bounds of familiarity, in shadowy, interstitial spaces shrouded in mystery; as such, these interstitial spaces are also home to radical utopian possibility, to the potential for new ways of living and being to be born that challenge and transgress established norms.

Ecotopia Rising

The critical utopia highlighted in Chapter 3 resists the exploitation and domination of people and the natural world increasingly characteristic of Western-capitalist societies, whilst evincing a fervent desire for mutual aid, ecological resilience, liberation, and peaceful living (Moylan 1986, p. 11). Critical utopias also, crucially, warn of the dangers of imposing rigid blueprints that bulldoze over life's complexity, casting their critical gaze outwards as well as inwards and highlighting the ongoing tensions that always attend intraspecies cohabitation and world-building. Many of the texts thus surveyed to varying degrees exhibit these characteristics. Relatedly, a common thread seen especially in Robinson's *Pacific Edge*, Le Guin's *Always Coming Home* and Starhawk's *The Fifth Sacred Thing* is the notion that perfection and total isolation are untenable myths; no ecotopian experiment can remain static or separate from dystopian forces without and within for long. 'The better', like reality, is always a continual striving requiring constant (re)negotiation, with no teleological 'end' point. The texts examined can be classed as 'ecotopias' in their varied denunciation of exploitative human relations with terra and other species, and imaginative reconstitution of these relations along more ethical, egalitarian and respectful lines. The 'ecotopian methodology' of permaculture, an 'ecological design science grounded in a fundamental recognition that economic viability and social justice are interrelated with functioning ecological systems' (Lockyer & Veteto 2013, p. 11), also variedly features in the design of the ecotopian societies depicted throughout. Similarly prominent are permaculture's key ethical principles – 'earth care, people care, and fair share' (Lockyer & Veteto 2013, p. 12), wherein each text foregrounds the important link between egalitarianism and socio-ecological wellbeing. Poignant reflections on intraspecies loss, love and life abound. However, there is variation in terms of the extent to which other terrestrials are depicted as central, active persons in the works, with *The Fifth Sacred Thing* and *Always Coming Home* as most closely approximating a 'terrestrial ecotopianism' rooted in terra and with other earth kin. It isn't enough to focus on ecological resilience without active care for, and attention to, the singular needs and lifeways of our co-terrestrials. By learning to listen more carefully and keenly, paying closer attention and becoming alive to the narrative agencies around us, we can ensure that ecotopian strivings for the 'good life' remain inclusive of our other-than-human kin.

CHAPTER 5

'The Great Refusal': Ecotopian Social Movements

> True, the dangers threating humanity [along with our terrestrial kin and terra] have in no way diminished, but the number of people who are aware of them is rapidly increasing. The dark clouds are threatening, coming closer and closer, but they have a silver lining which, though narrow enough, is increasing in luminescence
>
> – Lorenz (1973, p. xi)

In a sense most of today's progressive social movements can be designated as 'utopian' to varying degrees by way of their fervent mobilisations against a deficient present and ardent affirmation of the possibility of better worlds. Movements like the Zapatistas, Occupy Wall Street, and more recently the Sunrise Movement and Extinction Rebellion, all began by crying a resounding 'No!' to austerity and bank bailouts, wars, fracking, pipelines and oil subsidies, to the theft of land from local and indigenous communities, the privatisation of water and other vital resources, and to the increasing concentration of wealth in the hands of a tiny global minority. Rather than casting their sights to distant times and places in search of new colonial frontiers, as evinced by the post-terrestrial utopianism of the techno-capitalist elite, these movements have sought to build a better world for all in the here and now. For instance, the Sunrise Movement is a youth-led grassroots political movement 'united in the shared fight to make real the promise of a society that works for everyone' (Sunrise Movement 2021). Specifically, they seek to combat the climate and socioeconomic crisis through a Green New Deal – a just transition across every sector of society such as housing, transport, and agriculture towards clean and renewable energy production while providing millions of well-paying jobs in the process. Thousands have mobilised across the US in non-violent modes of activism, including occupying offices in the US Capitol, to push the Green New Deal onto the mainstream political stage.

My own research and interests have centred around radical environmental movements and activists, who more than most tend to place the rights and wellbeing of our terrestrial kin at the forefront of their visions and mobilisations for better worlds. And, as the focus of this book is on contemporary modes of ecotopianism, it is to such groups that I turn in this chapter. Below I explore their unique origins, ecological worldviews, tactics and how their deep kinships bonds with other terrestrials influence their distinctive modality of ecotopianism. I draw largely on semi-structured interviews (Aberbach & Rockman 2002) conducted with REAs from an array of movements and groups (i.e. Earth first!, Sea Shepherd, Hambacher Forst, Extinction Rebellion) during my PhD research between August 2017 and May 2019, as well as document and thematic analysis (Braun & Clarke 2013) of more recent mobilisations such as Ende Gelände, JSO, and Les Soulèvements de la Terre. Specifically, I was interested in why REAs risk life and limb for the protection of the Earth and its myriad inhabitants, rather than adopt a 'slow snail-like' move towards sustainability (Latour 1993, p. 5). Another motivation stemmed from my colleague Helen Kopnina's (2012, p. 236) observation that '[t]hose committed to the struggle of "radical environmentalism" or animal liberation are among the least understood of all contemporary opposition movements, not only in tactical terms but also ethically and philosophically' (p. 236). As such, I sought to critically investigate some of these key dimensions, namely: what motivates their engagement in often high-risk direct-action tactics? How do their post-anthropocentric ecological worldviews inform their activism? And, crucially, what are REAs' ecotopian visions for a more liveable and socio-ecologically just futures?

On the History, Nature and Tactics of REA Movements

Haunted by the spectre of nuclear war, 'post-war environmentalism' emerged marked by a burgeoning awareness of the global ecological ramifications of rampant technological change and socioeconomic activity within industrial society (Cassegård & Thorn 2018, p. 565). This awakening proliferated across Western Europe, the US, and Australia like wildfire, featuring campaigns for clean air and water, the safe disposal of sewage and waste, and improved public health (Rootes 2014). Modern environmentalism in the Global North[1] was effectively born when over 20 million Americans flocked to attend the first Earth Day on 22 April 1970 (Garcia-Hernandez 2007, p. 295). Throughout the 1970s, a series of key developments – mounting scientific evidence of environmental degradation, world reports on the limits to growth, the establishment of the UNEP, and crucially the 'political space' opened up by the New Left, anti-nuclear peace and countercultural movements of the late 1960s and early 1970s – helped entrench 'reform' environmentalism as the new dominant environmental discourse (Meadows et al. 1972; Rootes 2014). Environmental movement organisations (EMOs) such as Greenpeace and Friends of the Earth (FoE) gradually became institutionalised, accepted by governments as authoritative actors and sources of key environmental expertise, and experienced rapid growths in membership and resources. Moreover, they underwent increasing professionalisation which, crucially, was accompanied by shifts in repertoires from direct-action and protest towards conventional tactics such as lobbying[2] (Van der Heijden 2006; Berny & Rootes 2018, p. 949). Seeking cooperation with, rather than opposition to, government and industry, reformist environmental

---

1   A fascinating 'Global South' precursor to the proliferation of direct-action in the West is India's Chipko movement, founded in 1973 in Uttar Pradesh, wherein rural and indigenous people fought to reclaim control of the forests on which their livelihoods depend from commercial interests and colluding bureaucrats who sought to exploit the forest's resources by chaining their very bodies to trees in order to prevent them from being felled (Haynes 1999, p. 227). And of course, resistance by local and indigenous groups in the Global to extractive and mega infrastructural projects continues apace, as with the committed local resistance across India to the Sardar Sardovar mega-dam projects throughout the early 90s which managed to force the World to withdraw its support (Roy 1999; Klein 2015).
2   Indeed, recent anti-fracking and fossil fuel divestment campaigns largely adopt reformist emphases on legislative change and other parliamentary modes of political activity (Ogrodnik & Staggenborg 2016).

organisations embraced sustainable development or 'ecological modernization' narratives which maintain that there is no inherent incongruity between boundless economic growth and biospheric integrity (Andersen & Massa 2000). Rather, the aim became to reconcile economic growth with environmental sustainability through a 'greening' of global capitalism, that is, through technological innovation and individual consumptive changes. For emerging radical segments, however, addressing the roots of the climate and ecological crisis required no mere technological or legal alterations but fundamental transformations in our dominant socioeconomic systems, values, and worldviews.

Radical Environmentalism

From around the mid-1970s, growing disillusionment amongst more radical segments with the perceived philosophical and political shortcomings of liberal reformist environmentalism spurred a new 'fourth wave' of environmentalism known as 'political ecologism' or 'radical environmentalism' (Rootes 2014). Groups such as Earth First! (EF!), Sea Shepherd Conservation Society (SSCS), Hambacher Forst, the Climate Camp movement and to an extent the new civil disobedience movement Extinction Rebellion marked a 'step change' within the history of environmental mobilisations (Rootes 2008, p. 613). As discussed, reformist environmental groups often deploy tactics for curbing ecological decline that are deemed compatible with Western liberal-democratic structures and which cause minimal disruptions to the economic status quo. Pellow's (2014) helpful three-part characterisation scheme sheds light on the political, tactical and ideological distinctions within and between environmental movements, i.e. (1) 'radicals' who seek to entirely replace existing political-economic systems, (2) progressives,[3] and (3) mainstream

3   Pellow (2014) characterises SSCS as a progressive rather than radical group in light of its recent shift towards cooperating with state and corporate actors, and defining itself as a global 'anti-poaching organization dedicated to enforcing and operating

groups like the Humane Society who work towards incremental changes entirely from within the system (p. 57). Following Pellow (2014), reformist or 'progressive' environmental movements – a category that can be said to include XR – generally seek changes from *within* the present system by demanding that governments implement necessary legal and socioeconomic changes for mitigating climate and ecological breakdown. Of course, in actuality there are no rigid reformist/radical binaries; productive interchanges have always existed between more 'radical' and 'reformist' ends of the environmental movement spectrum. Indeed, as Malm (2021) points out, this productive disjuncture is precisely how progressive change occurs through what's known as the 'radical flank effect'. By engaging in more radical tactics and demands, radical movements make the demands by more mainstream organisations and movements appear more reasonable and palatable in comparison.

REAs, on the other hand, level multidimensional systemic critiques against the socio-ecological depredations of modern capitalism, specifically its need to continuously expand, extract and commodify people and places in its insatiable pursuit of shareholder profit maximisation. Moreover, they are also often highly critical of Western anthropocentric assumptions of human superiority to and separateness from other terrestrials and terra, and the consequent belief that the latter are ours to do with as we wish (Alberro 2020). For REAs it is oppressive systems like capitalism and Anthropocentrism that must be wholly dismantled; tinkering around the edges through, for instance, funding renewable energy schemes, amounts to too little, too late (Hernandez 2007, p. 296). Another key feature of radical environmentalism has been an especially strong undercurrent of apocalypticism, wherein widespread ecological collapse is seen as not merely

---

within the bounds of international environmental law' (Bondaroff 2020; Milstein et al. 2020). However, unlike mainstream litigation-based environmental movement organisations (such as Greenpeace and the WWF), SSCS activists generally exhibit more 'radicalised' identities marked by a rejection of 'involvement of mainstream society and with it "normative" methods for bringing about social change' (Stuart et al. 2013). Moreover, Sea Shepherd's co-founder Captain Paul Watson recently resigned from Sea Shepherd USA over their perceived institutionalisation and shift towards less 'radical' tactical repertoires (Kanowsky 2022).

on the horizon but already underway. This engenders a host of negative emotions such as grief, anger and a strong temporal sense of urgency, which in turn serve as powerful action motivators (McNeish 2017; Cassegård & Thorn 2018; Alberro 2021). As such, more recently some have delineated the emergence of a 'postapocalyptic environmentalism' (Cassegård & Thorn 2018; Cassegård 2023) wherein catastrophic environmental loss is seen well underway and inevitable, yet wherein there is still scope to salvage many species and ecosystems that still remain. The participants discussed at length below indeed exhibit a powerful 'postapocalyptic' desire to protect what remains amidst mounting tides of ecological collapse. However, more than this, many seem driven by a concrete utopian conviction that not all hope is lost, that another world more conducive to the flourishing of life amid and after the wreckage is possible, and that it ought to be fought for in the here and now (Alberro 2021).

Disrupting the Present

REAs engage in a 'politics of contention' that involves trenchant criticisms of the structural drivers of ecological despoliation and, notably, the use of direct-action repertoires designed to disrupt, slow, and ultimately reverse climate and biodiversity collapse (Tarrow 2013; Tilly & Tarrow 2015). Direct Action (henceforth DA) was originally coined by anarcho-feminist activist Voltarine de Cleyre (1912). However, this subversive mode of activism received its formal theorisation in the wake of Gandhi's opposition to Apartheid in South Africa and British colonial rule in India, and further evolved from the non-violent direct-action strategies of civil rights and anti-war demonstrations of the 1960s and 1970s (Plows et al. 2004; Tarrow 2013). DA entails a specific kind of political action associated with anarchist praxis that is 'not beholden to the political tactics of the liberal state' (Heynen & Van Sant 2015, p. 173). DA takes a number of different forms, for example, civil disobedience, road blockades, 'climate camps' (Saunders 2012), and notably, the roving transnational 'conflict zones' of local and indigenous resistance against extractive industries – from oil pipelines to logging and mining operations – termed 'Blockadia' (Klein 2015; Estes 2019). More controversial forms of DA are those that include acts of economic sabotage or 'ecotage' such as arson and the dismantling of ecologically destructive machinery. In response the latter, many have drawn on the pejorative sense of 'radical' and referred to REAs variously as 'extremists' and even 'eco-terrorists'[4] (Jarboe 2002; Liddick 2006; Nagtzaam & Lentini 2007; Eilstrup-Sangiovanni & Bondaroff 2014; da Silva 2020). Of course, REAs are launching fundamental and multidimensional challenges to Western-capitalist worldviews, institutions, productive and consumptive modalities, embodying what Argentine philosopher Enrique Dussel (1985) refers to as an 'exteriority' which avows the possibility of a radically different world (p. 46). Therefore, such mobilisations for total liberation can only ever be illegal as they seek to resist unjust institutions and legal systems themselves.

Human ecologist and activist Andreas Malm (2021) asserts that blowing up a pipeline is a categorically distinct form of violence compared to killing people, and likens their conflation to 'conceptual abuse' (p. 108). Rather, Malm suggests that, with the world fast on track to exceed 2 degrees of warming by 2100 due to the 'wanton criminality' of the fossil economy which is driving anthropogenic climate change and killing people, this 'only underscores the need for emergency tactics' (p. 151). Drawing on previous scholarship on political extremism (Safire 1996; Midlarsky 2011), Bötticher (2017) makes a further crucial distinction between political 'extremism' as opposed to 'radicalism'. The former refers to an ideological position embraced by dogmatic anti-establishment movements which typically: (1) view politics as a struggle for supremacy, (2) are willing to engage in – and at times glorify – criminal acts and mass violence in order to gain political power, (3) eschew diversity in favour of social homogenisation and (4) exhibit a penchant for totalitarian rule over democratic modes of political participation (p. 74). REAs espouse anti-system or radicalised identities

---

4   The term 'eco-terrorism' was coined by former leader of the Center for the Defense of Free Enterprise (CDFE), Ron Arnold, in a 1983 article entitled 'Eco-Terrorism' published in Reason magazine (Smith 2008, p. 545).

via their engagement in extra-parliamentary modes of political activity for effecting fundamental transformations in mainstream socioeconomic and political systems (Dalton & Rohrschneider 2003). However, REAs venerate diversity and the irreducible singularity of the other – indeed, many are mobilising against the systematic annihilation of life's diverse forms amid the Capitalocene. Moreover, they overwhelmingly favour direct-democratic and non-hierarchical political modalities, and thoroughly reject the use of violence against living entities as a political tool. In stark contrast, radicalism is an emancipatory sensibility and strategy that seeks to enact sweeping socioeconomic and political transformations for the betterment of all (Bötticher 2017). REAs and similar groups are thus categorically distinct in aims and tactics to far-right mobilisations, for instance, which often deploy violence against people and thus more accurately fit the label of 'extremist' (Youngblood 2020). Lastly, it must be reiterated that the worldviews and actions of contemporary REAs are far from new. As discussed in the following chapter, Indigenous peoples the world over continue to engage daily in fervent resistance against (neo) colonial-capitalist extractivism that threatens their very existence (Estes 2019; McGregor et al. 2020). Nevertheless, what drew my interest as a researcher is what motivates and sustains environmental resistance movements emerging from 'within the belly of the beast' (Demaria & Kothari 2022, p. 143) – i.e. Western-capitalist societies – when their wider society seems so antithetical to the worlds they wish to bring into being.

Insurgent Realities: Crying '*No*' to the Existing Order

Cathcart's (1978) distinction between 'confrontational forms' (radical) as opposed to 'managerial forms' (reformist/progressive) of social movements adeptly captures the utopian alterity of REAs. The latter, as elucidated by mainstream ENGOs, deploy 'managerial rhetorics' designed to portray – or at least take for granted – the existing system as *the* true order on the historical trajectory towards a more civilised state. Thus, such approaches rarely if ever question the underlying onto-epistemological foundations or ethico-political orientations of the status quo within which they find themselves. By contrast, the *confrontational* rhetoric' such as that deployed by REAs[5] surfaces during periods of profound societal (and ecological) breakdown, or when society's moral underpinnings are being called into question (Cathcart 1978, pp. 238, 242). Confrontational movements embody 'a kind of ritual conflict whose most distinguishing form is confrontation with the established order, its structural and ideological hierarchies, and its values' (Cathcart 1978, pp. 235, 237) (also a distinguishing feature of utopianism; refer to Chapter Three). In one sense, all political movements can be said to be born when the multitudes say 'No!' to the 'existing structures and politics of the dominant society' (Breines 1989, p. 23). However, this is even truer of the confrontational forms, which position themselves as profoundly alienated from the prevailing order (estrangement as yet another key feature of utopianism). Such millenarian movement forms are 'moved by an impious dream of a mythic New Order – ... to rise up and cry *No* to the existing order ... saying, "You cannot go on assuming you are the true and correct order ..."' (Cathcart 1978, p. 243) and 'prophesy the coming of the new' (Griffin 1969, p. 460). Indeed, there is no single 'true and correct' order, but many possible forms, and some far more conducive to multispecies flourishing than others.

Cathcart goes so far as to suggest that many of the so-called 'types' of movements described in recent literature do not appear to be movements at all, but 'adjustments to the existing order' rather than rejections of it (1978, p. 239). The significance of the confrontational movement lies in its ability to engender the realisation that the conflict is no mere re-alignment of the established order but a *fundamental challenge* to it, a striving for deconstruction and reconstruction rather than mere reform. In other words, following Audre Lorde (2017), 'The master's tools will never dismantle the master's house. They may allow us temporarily to beat him at his own

---

5   According to Cathcart (1978), the proclamation of an alternate vision is where movements truly begin, because 'up to the point of confrontation it is impossible to know that a radical or true movement exists' (p. 241). In other words, confrontation engenders and reinforces an emerging movement's identity, substance and form, and thus they cannot be movements for radical change without it (p. 243).

game, but they will never enable us to bring about genuine change' (p. 17). When the house itself is out of order, a better home can only be (re)built with radical tools – i.e. love, rage and hope (Alberro 2024) – that respect and cherish difference and diversity, deploying them as essential building blocks for the worlds we wish to bring into being. Hence the etymological sense of the word radical, denoting *roots*. Confrontational movements, like prefigurative modes of political action, are thoroughly utopian in that they seek to facilitate *systemic* change. Their mobilisations and sensibilities emerge as pockets of radical difference and negation in defiance of oppressive systems and structures, and embody glimmers of a potentially better 'Not-Yet'. Traditional research on social movements has to an extent tended to neglect more utopian or 'prefigurative' movements (Jeffrey & Dyson 2021) of the periphery that seek to supplant existing socioeconomic systems and modes of relationality with better alternatives. The spaces of alterity that REAs embody and seek to instantiate, serve as 'parallel universes' (Klein 2002, p. xxiv) wherein the dominant norms of our individualistic, hypercompetitive, and ecologically rapacious socioeconomic systems and institutions lose their facades of immutability through their juxtaposition with arguably better alternatives. It is this that my research with REAs has sought to shed further light on, in particular the content of their visions for more socio-ecologically just futures.

REAs' Post-anthropocentric Sensibilities

One particularly notable experience comes to mind that speaks considerably louder than much of what I could articulate about my experiences with REAs. Earth First! (henceforth EF!) was born in 1980 in the Southwestern United States (Wall 2002) and has since become one of the most widely known anarchic-style environmental direct-action activist groups in the English-speaking world (Rootes 2015, p. 422). During my PhD research, I'd been attending the EF! 2017 summer gathering, when the second day had drawn to a close after a series of thought-provoking workshops, action planning, practical skills training and forays into the worlds of green speculative fiction. I had decided to try one of the ciders from the camp's makeshift bar, the minor proceeds being directed entirely towards funding the event and the Earth First! collective. After receiving my pint, I was about to take the first sip when I noticed that a spider had fallen in and appeared to have drowned. After painstakingly searching for signs of life I concluded, somewhat prematurely, that it was too late for the unfortunate creature. As I sought to remove him/her from the glass, one of the activists intervened and determinedly attempted to revive him/her. This person's thoughtful efforts proved successful and the spider ultimately regained consciousness. Such efforts proved all the more incredible in light of the fact that the creature in question was one that many perhaps wouldn't give much consideration to, indeed, are often thoroughly hostile towards. This incident offers a telling example of REAs' 'post-anthropocentric' (Braidotti 2013; Ferrando 2016) worldviews, which feature staunch repudiations of notions of human supremacy, and emphasise the inherent value of – and deep kinship bonds with – other terrestrials. However, as discussed below, variation can be detected in their ecological worldviews and modes of interspecies relationality, which constitute a mosaic of environmental-ethical approaches – from biocentric valuations of life itself (Rolston 2020) to ecocentric emphases on the value of wider biospheric integrity (Alberro 2020).

In EF! communications, art, diagnostic and prognostic narratives, other species as agentic kin feature far more prominently than most other groups investigated. Hence the name 'Earth First!', a hierarchical reversal that in the movement's earlier years took on misanthropic tones. However, in today's iterations, we see less a reversal of traditional human/other-than-human hierarchies and greater efforts to dismantle hierarchical systems and relations altogether. Just Stop Oil (JSO), one of the new A22 movements, emerged in the UK in 2022 with the aims of deploying civil-disobedience-style direct-action tactics designed to pressure the UK government into halting investments in, and the licensing and production of new fossil fuel projects.[6] It has since become an 'international network' with chapters in

6   Curiously, JSO is heavily funded by the Climate Emergency Fund, which has received sizeable donations from Aileen Getty, granddaughter of oil tycoon J. Paul Getty.

the US, Norway, France, Australia and other Global North countries. By contrast to EF!, JSO's direct references to 'nature' and other species are sparse. 'Gaia's Navy' (Nagtzaam 2013), or the SSCS, was initially conceived as the 'Earthforce Society' in Vancouver, Canada in 1977 by its founder, Captain Paul Watson (SSCS, 2016), a former Greenpeace co-founder who was sacked from the organisation for his vigilante-style direct-action tactics (Watson 1982, p. 153). With chapters across the world, SSCS and their fleet of twelve ships[7] managed to accomplish what decades of formal international law enforcement mechanisms and litigation by more established environmental NGOs have often failed to do: significantly reduce the slaughter of whales and other marine life. Notable direct-action feats included positioning their vessels between whales and harpoon ships, ramming whaling vessels in order to prevent harpooning and refuelling (Nagtazaam 2013), and ecological ecotage of equipment and property used to harm or kill wildlife (Eilstrup-Sangiovanni & Bondaroff 2014). However, more recently SSCS have begun to focus on strategic engagements with institutional means and state actors, calling attention to the legal violations of whalers and citing the global moratorium on commercial whaling instituted in 1986 by the International Whaling Commission (IWC). SSCS now defines itself as a global 'anti-poaching organization dedicated to operating within the bounds of international law' (Watson in SSCS, 2019). The organisation's growing emphasis on cooperation *with* states and relevant organisations (Milstein et al. 2020) led Captain Paul Watson to resign in 2022 (Kanowsky 2022) in protest over Sea Shepherd USA's shift towards reform over radical action. However, whither the other-than-human in SSCS activist worldviews? Below are two indicative examples:

> I said to myself, 'Here we are killing this incredibly intelligent, beautiful, self-aware, sentient being for the purpose of making a weapon meant for the mass extermination

---

7   In addition to donations from their numerous supporters worldwide, SSCS receive generous sums from their wealthier and more high-profile supporters such as former American TV show host, Bob Barker (after whom they named one of their vessels). This presents a curious incongruity between their 'radical' and in some instances, anti-status-quo demeanour whilst also being funded by powerful capitalist actors.

of human beings,' and that's when it struck me, as a species we're insane .... So I developed this sort of a biocentric point of view, and I've been fighting against this anthropocentric-dominated culture ever since. (Captain)

They're [whales] ecologically so important and so valuable that, even when they die, they sink to the ocean and other lives, especially deep-sea organisms, can live on them. So, they have such a crucial role, and therefore, it makes absolute sense that we say, 'This life is more important than this life.' (Shark)

Captain calls on others to be concerned for the plight of the whale, exemplar of the charismatic megafauna that has captivated the public imaginary, because they are 'intelligent, beautiful, and self-aware'. However, recalling the discussion in Chapter 2, what of the widescale and rapid decimation of insect populations across the globe? The aforementioned, still infused by aesthetic-based valuation (Rolston 2002), harbour inklings of an ethic and 'ontology predicated on affection for sameness' (Sargisson 2000, p. 146). Similarly, Shark's prioritisation of whales due to their purportedly significant ecological roles still exemplifies a hierarchical valuation system wherein species are ranked according to the value of their ecological functions. Though such post-anthropocentric views might be critical of human exceptionalism, they fall short of being truly transgressive (Sargisson 2000, p. 146) in the sense of dismantling antagonistic dualisms and hierarchies altogether. However, the excerpts below are indicative of a much more common thread, wherein most REAs strove to deconstruct hierarchical and dualistic constructions of 'nonhuman' otherness (Plumwood 2002; Latour 2018):

> For some reason we've separated the world into these three spheres where there's nature, there's animals, and there's us, and it's one of the most arrogant, kind of, separations in the world .... So I think we need to reinvent that [notions of human separateness from the natural world] ... we are animals. (Jellyfish)

> We've become very nervous-system[-focused]. Like, a lot of Animal Rights folk won't give credence or any time to any kind of theorizing or philosophizing, or experiential musings on the fact that plants, trees, etc. might have a degree of being, or sentience and intelligence that we can't comprehend as yet because they lack a central nervous system .... It's this grading of superiority. (Badger)

Jellyfish and Badger variedly shed critical light on the rigid hyper-separation between humans and other terrestrials (Plumwood 2002; Latour 2004), which has elevated the former to a superior status while the latter has historically been defined as 'lack' in relation to the Master (human) identity (Plumwood 2002a). REAs' post-anthropocentric sensibilities translate into deep kinship bonds with other species and terra, serving as powerful action motivators fuelling their engagement in often high-risk tactics such as tree-sits, road blockades and, in the earlier years of Sea Shepherd mobilisations, collisions with whaling vessels. These kinship ties result in their feeling so emotionally and socially 'at stake' in the lives of earth kin that their loss is experienced as the *'severing of a social bond'* (van Dooren 2014, p. 136), resulting in powerful emotional and psychological experiences of grief (Cunsolo & Ellis 2018):

> In my own lifetime in what is one of the most privileged countries in the world, I've also seen the creep of the absence of life into this country, this island. So, hedgehogs, badgers, foxes, the birdlife … all these factors, again, they weigh on me … (Badger)

This phenomenon is akin to but goes beyond 'Solastalgia', the term coined by philosopher Glenn Albrecht (2020) in reference to 'the pain or distress (i.e. psychological, emotional) caused by the lived experience of ongoing loss of solace and the sense of desolation connected to the present state of one's home and territory' (p. 11). In the context of REAs' lived experiences of the loss of other terrestrials, the incalculable grief arising therefrom is not merely over the loss of one's home but, rather, from the irreversible severing of kinship bonds with cherished kin. Although, perhaps other species as kin are seen as offering a sense of solace, thus the term is not too distant, particularly in relation to Albrecht's related term 'soliphilia' as that which motivates the political commitment and collaboration 'for protecting and conserving those places [and beings] that we love' (2020, p. 21). As I'll explore below, REAs' deep kinship bonds to Earth others and grief over their loss redoubles their commitment to preventing further losses and salvaging those who remain. It also (re)configures their relations to utopianism, hope and notions of futurity in complex ways.

On REAs' Terrestrial Ecotopianism

The Great Refusal

As we saw in Chapter 3, utopias' transformative capacity stems from their dual processes of negation (of a deficient present) and annunciation (of better alternatives). There is a distinctive 'nowness' to REAs' staunch critiques of and resistance to the present (Moylan 1986). Moreover, many exhibit a considerable degree of postmodern scepticism toward closure around particular visions of 'the good' (Foucault & Miskowiec 1986; Levitas & Sargisson 2003, p. 15). This is largely a result of their apocalyptic portrayals of extant ecological collapse (McNeish 2017) in light of the multiple ecological perturbations that proliferate amid the Capitalocene, which further complicate REA relations to hope and futurity by rendering inklings of a predictably delineable future increasingly opaque (Head 2016; Garforth 2018). Before going on to discuss REAs' particular modality of ecotopianism, it's worth examining what exactly they're mobilising *against* by elucidating their core critiques of the 'Now'. A central diagnostic theme that has prevailed throughout REA discourses since their earliest mobilisations are the purportedly multifarious socio-ecological contradictions of advanced industrial society (Manes 1990; Taylor 2008; Saunders 2012; Marcuse 2013). Most activists interviewed as well as the texts under analysis articulated variants of such critiques. For instance, the novel French radical ecological direct-action network Les Soulèvements de la Terre ('Earth Uprisings') seeks to dismantle the 'agro-industrial complex' and related extractive infrastructures and enterprises, as well as form the foundations of a new land recovery movement:

> The unbridled development of transport infrastructure is one of the major causes of the devastation: global warming, destruction of life ... Go faster, always faster. Constantly densify the flows that run the mega-machine. Extend everywhere the tentacles of the metropolis. (Les Soulèvements de la Terre 2023)

However, recently many REA socioeconomic critiques have taken on a more explicitly anti-capitalist configuration. They and many of the ecotopian texts discussed in the previous chapter thereby designate not only rampant industrialisation and growth-oriented socioeconomic systems more generally, as well as attendant over-consumption/over-exploitation, but especially capitalism (particularly in its late neoliberal modality) as a key driver of the current socio-ecological crisis (Harvey 2007; Strauss 2010; Foster et al. 2011; Žižek 2011; Braidotti 2013; Marcuse 2013; Klein 2015). REAs cite many of the reasons discussed in previous chapters. Consider the following exchange with an EF! activist who noted emphatically the predominant role that contemporary capitalism – and in turn transnational corporations operating under its logic – has in shaping our present predicament:

> The main contributing factor is called capitalism, and there is no doubt about that. I'm not going to pinpoint on who burns more CO_2, who produces more plastic, who kills more animals, umm, all of that is just a symptom of the illness, and that illness is capitalism without a doubt. When greed is over everything, including ethics, and when we become so greedy that we can't see further away than ten-years'-time and even five-years'-time, that is a symptom of capitalism ... Obviously, I wouldn't say that capitalism is good but, through capitalism we've had a lot of technological advancements, and it is probably now the time to detach ourselves from a rotten system that is destroying the whole world and oppressing human and nonhuman animals, as well as the whole environment ... (Koala)

Many participants made similar references to capitalism as an insatiable, endless-growth-oriented system that corrodes ethical relations by placing shareholder profit maximisation above all else:

> ... this idea that we have infinite resources, and going back more into capitalism, this idea that profit is a good thing, that expansion and growth is a good thing ... It's that capitalism has to expand in order to survive, GDP, blah, blah, blah, and we realize that we live on a finite planet, there's an automatic contradiction there. You can't expand infinitely on a finite planet ... (Bear)

> we live in ... a global capitalist economy now, that believes in ... infinite growth economies that have to have infinite growth to survive and to function ... (Warrior).

This of course is not a novel observation but influenced in part by limits-to-growth discourses of the 1970s (Meadows et al. 1972) mentioned previously. However, contemporary crises – climate change, rampant deforestation, mounting scarcity of rare earth metals, fresh water, etc., pollution and toxification – as the inevitable consequences of a socio-economic model predicated on the maxim of endless expansion and acquisition render such diagnoses especially pertinent. For those such as Foster et al. (2011) and Clark and York (2005), and posthuman scholars such as Braidotti (2013), capitalism is inherently and irremediably incongruous with ecological integrity. Boundless accumulation necessarily proceeds via the commodification of 'nature', prioritisation of exchange values over use values, the alienation of humans from their essential material foundations, from one another and from other terrestrials. Other REAs echoed similar concerns, especially problematising capitalism's post-1980s neoliberal manifestation:

> ... the problem is that this rampant neoliberal form of capitalism that we have is so driven by the top 1% having 99% of the wealth, and doing essentially whatever they like to create profit for themselves and screwing over everybody else, is what I would count as the main driver. (Jellyfish)

> there's been this creep towards a sort of neoliberal consensus where (A) the money's funnelled to the people at the top of society, you get privatization, you get raiding of the pension pots, you get the deregulation of an economy so that a small country in the Global South has to open up its financial flows so that big investors can pile the money in and they can just move their money around, and capital is quite mobile, and (B) there's also this underlying political mantra where growth is necessary and is good, and I've yet to see a government win on a de-growth ticket or even on a sort of zero-growth ticket. I think these are the kinds of things that are driving the ecological destruction. (Squirrel)

> Yeah, the main driver is money. We have multi-billion-pound industries, especially the fossil fuel, that are run by narcissists and psychopaths, or sociopaths. Umm, they live in their bubble, and the only way for them is just growth, growth, growth, and they are trying to extract infinite resources from a finite planet, and it just obviously doesn't equate. (Butterfly)

For Meadow, the particular damage wrought by neoliberal capitalism stems from its perpetuation of the idea that profits ought to be the sole raison d'être of socioeconomic activity:

neoliberal economics that have actually, really, massively increased this idea that a company's sole purpose is to keep increasing profits to its shareholders, and everything else just gets pushed aside.

A related diagnostic framing subtheme, and one which relates to REAs' general disavowal of ownership and possession, is a mode of being prevalent amidst Western industrial capitalism alluded to previously, which entails not only the insatiable desire to grow and consume but more fundamentally the subordination of being to having. The 'being' mode, as described by Erich Fromm, involves being '*oned* to the world' and is characterised by interconnectedness with other beings and our wider natural support systems (Fromm 1976). The 'having' mode, on the other hand, is an instrumental economic rationality (Gorz 2010), ethos and orientation predicated on the avaricious acquisition of things, profit and power. The 'having modality' rests upon a profound alienation from others as mere things to be acquired or competed against. Fromm saw this modality as especially prevalent in post-war, mass consumptive Western-capitalist societies (Fromm 1967, p. 57), and thus a key driver of contemporary socio-ecological decline:

> this weird mentality where we've always got to get the next best thing, and we're making massive landfills, we're taking finite resources just to make more crap that we really don't need. We have everything already. These developing nations, they want everything that we've got and they have some of it already, and there just isn't enough resources on the planet for this mentality that we've all got to have TVs, cars, and god knows what. (Poseidon)

As discussed in the previous chapter, this was also a palpable diagnostic theme in Huxley's *Island*. For many activists, the depredations of global capital intersect with and exacerbate a multiplicity of other oppressive and exploitative modes, which is why no amount of 'greening' or constraints placed upon its inherently self-destructive tendencies will be sufficient:

> Capitalism hasn't just brought about environmental destruction, it's also brought about lots of other things like oppression of women, oppression of children, oppression of outside groups, the third world or Global South. So, I want capitalism to end, not just for the environment, although that is the most pressing issue, but like,

as long as we have capitalism we'll still have inequality, and hierarchy, and so on. *It all has to go.* [emphasis added] (Fox)

Think of capitalism and the current industrial civilization that perpetuates it as **a massive gnarled tree**; it's very easy to atomize our grievances into single issue campaigns that don't share a common strategic goal. One campaign tirelessly hacking at one branch, another campaign tirelessly hacking at another – but **the trunk continues to stand strong and continues to sprout new branches**. (EF! 2017, p. 30)

As with Bookchin's (1986) organismic metaphor of capitalism as a cancerous scourge that permeates and corrupts all, the interview extracts above suggest perceptions of capitalism as thoroughly incompatible with more socio-ecologically just modes of being. As Warrior denotes, 'You just can't have equality and capitalism at the same time, it's impossible'. Therefore, detectable amongst many REA diagnostic framings is a decidedly utopian desire for a thorough uprooting of oppressive systems for mitigating the socio-ecological crises that mark the Capitalocene. This is what differentiates REAs from more traditionally reformist environmental groups and movements, which tend to emphasise a tinkering around the edges of the current system and the unethical modes of being that it generates, rather than dismantling it entirely.

A final point worth noting is some REAs' allusions to the ambiguities surrounding culpability for Capitalocene woes. Meadow, for instance, views agency and culpability in relation to Capitalocene decline as not associated with one particular actor or group of identifiable aggressors, but rather as diffused across social systems, reproduced and reinforced in every interaction and transaction that we undertake (Latour 2017):

We're all, well, nearly all of us are bystanders, **and in our passivity we're actually active contributors. We're all consumers. We all other. We all deny.** You know, we all have a place within the matrix of oppression, you know, **we're all oppressed and we're all oppressors**. I think it's very easy to demonize individuals or corporations, or capitalism, but I would say all of those are emergent properties of a system. So, I would say, you know, **nobody is running capitalism**. You know, there are many hundreds of thousands of individuals with a larger stake in it than me, with more power than me, and so obviously they have more impact than me to maintain things to be cozy for them. But, fundamentally, **capitalism is an emergent property of billions of micro-decisions** that are made every day. You know, we're all complicit, we all do capitalism in our small way.

On the one hand Meadow alludes to the notion that many of us are bystanders, passively looking on as socio-ecological upheavals gather momentum; yet he also invites us to consider how this monolith known as 'capitalism', or the propensity to oppress and *other* others, isn't some alien thing *over yonder* that we (or at least most of us) can claim to be non-complicit in or extricate ourselves from. He echoes previous sentiments of a system or machine that is 'out of control' with no single entity in charge, while further emphasising that its nature and direction are dictated by billions of micro and macro-interactions. Meadow's response, in a manner similar to Morton (2010) and Latour (2011), gestures towards a recognition of how the intermingling of nature-cultures and inter-actant imbroglios obliterates distinctions between inside versus outside, background versus foreground, and perpetrator versus totally passive or innocent bystander. In a world characterised by entangled multiplicities of human and other terrestrial agencies who react to one another in complex and often unpredictable ways, separating one entity or group of entities and designating them as the source of all ills becomes increasingly dubious. Rather, amid the turbulent Capitalocene, 'there are no more spectators, because there is no shore that has not been mobilised in the drama of geo-history' (Latour 2017, p. 40). We are all implicated. However, make no mistake – BP, the US military-industrial complex, Elon Musk, and the like have a lot more to answer for than the average US citizen.

The Problem of Human Supremacy

Another key diagnostic theme is that of human-exceptionalist or 'anthropocentric' beliefs and worldviews, and the attendant speciesism that they foster. Particularly nefarious for REAs and many ecotopian texts is the quintessential anthropocentric conviction, characteristic of Fromm's 'having mode' (1967), that other species and the wider earth system are *for* us rather than *with* us (Le Guin 1985, p. 77):

> I have absolutely no more right than any other animal on the planet, and that we place ourselves as a superior species, that's the whole problem. We really just think that we

> are allowed to use, and use, and eat them, make service animals out of them whatever, and it's not true. What gives us more right to live than a cow or a whale? (Poseidon)
>
> Well, the main problem is anthropocentrism, which is *this attitude that we can take what we want, do what we want, everything else is expendable* [emphasis added], that we're the only important species that matters. (Captain)

In the previous chapter, we saw similar critiques surface throughout the ecotopian texts, wherein other species are often framed as purposive agents that are 'like the electric light ... they don't belong to anyone' (Huxley 2009, p. 15). In these instances, just or ethical treatment of other terrestrials is regarded as incompatible with the concept of ownership or any related exercise of domination over them (Cochrane 2009). Delfin further denotes that what's at issue is '[t]he belief that some lives are worth more than others', or the reduction of their value to that which serves our economic and other interests, which manifests in the language we use:

> We need a value system where trees are valued for what they are, for what they provide, not for what we can make out of them. We need a value system where we recognize that fish have to stay in the ocean to maintain ecological integrity of the ocean, they're more valuable in the ocean swimming around doing what they do than they are on our plates. (Captain)
>
> Economies are all based on continual growth, that success is growth ... and on a finite world, with finite, what we call, 'resources' ... that's the problem, *how dare we call other living beings resources* as if they are simply ours to exploit 'till they are gone? There's the root of it; we cannot continue treating other life on this planet as if it is something that has some economic value. It is worth *so* much more than that. (Stonehenge)

Stonehenge draws attention not only to the biogeophysical contradictions of the endless growth paradigm but, crucially, to the violence and profound arrogance of reducing our myriad and singular terrestrial kin to the status of mere fodder for fuelling the human enterprise. Though traces of hierarchical valuation on the basis of ecological function resurface in the excerpt by Captain, he similarly alludes to the ethical paucity of instrumental rationalities with regard to other terrestrials, wherein the latter are often valued only as means rather than ends in themselves. Such approaches, though exacerbated by the instrumental economic rationalities of neoliberal capitalism, are rooted in the longstanding tradition of anthropocentrism as an oppressive cultural dominant explored in Chapter 2.

A Better World Is Possible!

> At this gathering we are working towards a different kind of living-together in nature – non-hierarchically, low-impact and in solidarity and community. We want to challenge the multiple systems of oppression – patriarchy, white supremacy, transphobia, ableism, speciesism and others – and work towards new relationships with the earth, nonhuman nature and one another. (EF! UK Summer Gathering 2022)

REA visions of ideal socio-ecological relations often appear fragmented and contested. Much of their energies are often directed toward prefigurative enactments of 'the world we want now' rather than towards spatial-temporally displaced depictions of ecotopian structures and institutions. However, the apocalyptic narratives that pervade REA imaginaries are suffused with ecotopian desire, as embedded within portents of nightmarish worlds is a fervent desire (and hope) for the opposite: a world teeming with abundance (Sargent 1994; Garforth 2018). For instance, the following are indicative of a common theme throughout, wherein presages of doom are juxtaposed alongside allusions to the (utopian) potential for better worlds:

> Without people who will take a principled stand, tell it like it is and have a vision of a world which is not centered on the system that is destroying it, we are all doomed. Social collapse is already happening all over the world. The idea that we in the West can somehow wake up and change that is just another global spell. We actually have to engage with how we're going to face it, how we are going to live differently in these times when the planes will stop flying, and the lorries will stop delivering food…. How do we deal with this that is not in a way that is protectionist and exclusivist, and ridden with oppositional thinking? *There is another way.* [emphasis added] (Stonehenge)

Here Stonehenge is disrupting linear conceptions of time and historical progress: ecological collapse is not merely on the horizon but is already happening; we are already living in the 'end times' (Žižek 2011; Latour 2017). For many BIPOC people the world over, 'the house has been on

'The Great Refusal': Ecotopian Social Movements

fire for a long time' (Wretched of the Earth 2019). Yet, the infinite potentiality of other worlds has not been, indeed can never be, exhausted (Bloch 1986). Thus, the sense of finality encapsulated by notion of collapse refers to the 'present order' as it has materialised, not to the end of all possible worlds. Similarly, many REAs repudiate exclusively future-oriented modalities of hope:

> I'm also actively stepping away from the idea of hope, as it is traditionally espoused .... Hope projects something into the future, and right now I'm really focusing my energies on how to address how we live now, rather than how we might live in the future. (Stonehenge)

> I'm going through a phase where I actually don't have an awful lot of hope, and people go, 'Well, why do you continue doing what you're doing?' And if I had a garden where I had the last butterfly in my garden and I knew it was going to die, I would still do everything that I could to make sure that butterfly lasted as long as possible, you know? And so, it's just part of our makeup; it would be impossible not to look at our flora and fauna, us, and not want it to exist as it was, and as it should be. (Butterfly)

The perceived immediacy of ecological breakdown compels many REAs like Stonehenge to focus most of their energies on resisting the Now of widespread loss and disintegration. For them, an abstract, disembodied, predominantly future-oriented hope, or hope in the form of blind optimism (Treanor 2019), is a dangerous distraction. REAs' curious form of 'hopeless activism' evokes Duggan and Muñoz's (2009) 'critical modality of hope' (Alberro 2021), wherein hope and hopelessness are conceived not in an oppositional but rather in dialectical relation to one another. Herein the opposite state of hope is not hopelessness per se (Lazarus 1999) but complacency (Duggan & Muñoz 2009). When all hope fails, and being itself is perceived as meaningless, 'there is *nothing but despair* [emphasis added]', which can in turn morph into nihilistic complacency in the face of climate chaos and biological annihilation. Yet, REAs vehemently repudiate complacency and passivity; many are hopeless, indeed, but hopeless in the narrow sense of recovering specific loss – of cherished and departed kin, and of a benignly unfolding futurity (Lazarus 1999, pp. 654, 660). Many REAs continue to fight on behalf of those cherished Earth kin who are still under threat in the here and now, and for those who might yet branch forth:

> There's stuff today that is lost already, really, but there's also so much that's around that can be saved if we all put ourselves to the task .... It would be really, really sad to just give up now and say, 'Oh, it's all fucked, it's too late,' when actually, maybe it's not. I mean, it's definitely too late for a lot of things, but maybe for most things it's not. (Atacama)

Critical modalities of hope (Duggan & Muñoz 2009) move beyond complacency and denial, wherein grief – at widespread loss of life and of a delineable Not-Yet – is actively confronted (Head 2016) and utilised as a powerful action motivator. Atacama further alludes to the 'material openness' of reality, whose potentiality is never fully worked up – and wherein hope flourishes in its critical and concrete forms. These are to be distinguished from forms of optimism that proclaim the assured realizability of a particular desired end, therefore ascribing definite content to the 'To-Come' (Ware 2004). But, hope is possible precisely because we aren't certain of what's to come; certainty, like perfection, are 'sticks with which to beat the possible' (Solnit 2016). For many, it is indeed already too late, but the full potentiality of the To-Come is never exhausted – extant life forms and those yet to branch forth might yet thrive once again.

REAs' strained relationships to hope's more optimistic dimensions discussed may also result from attempts to situate oneself and come to grips with the event horizon of our *collective* 'ceasing to be' that Capitalocene crises portend (Haas 2016, p. 287). For Derrida (1976, p. 5) the 9/11 terrorist attacks constituted a true event in the sense that they were without precedent, thus shattering notions of an ordered world and benignly unfolding future whose horizon could be more or less predictably delineated (Wood, cited in Fritsch et al. 2018, p. 38). Similarly, Capitalocene crises effect such a 'disintegration of our horizons of significance and possibility', at times a 'complete dissolution of meaningful horizons' (Toadvine, cited in Fritsch et al. 2018, p. 57; Treanor 2018). For REAs, this manifests especially in widescale biological annihilation. Life in the form of extremophiles, for instance – organisms that thrive in places with extreme temperatures, anaerobic, and radioactive environments – will go on, as well as other organisms that adapt to the paroxysms and perturbations of an increasingly inhospitable world. But many of us and countless singular Earth others who are our co-evolutionary kin will likely no longer be. This is where the significance of grief and mourning over the actual and looming 'ceasing-to-be' of cherished Earth kin plays a central role in REAs' relationship to an ecotopian 'Not-Yet' and in motivating their continued strivings; what they mourn is the potential loss of an abundant future teeming with life in its myriad manifestations. Put another way, what REAs actively strive against – not only for themselves but crucially for others – is 'utopia' in its pejorative sense, the good place that is no place, in this instance the 'ultimate nowhere of non-being' (Clark in Davis & Kinna 2009, p. 9). In this vein, REAs frequently speak of an active engagement with and affirmation of grief, the departed, and those not-yet gone. This modality of mourning features an engagement with the externalised memory of the unsubstitutable 'other' and takes place via a future-oriented enactment of responsibility to the other (Kirkby 2006). One can posit REAs as engaged in a fervent resistance against a totally foreclosed futurity, in an ongoing effort to avow their responsibility to 'restore the other's futurity' (Diprose 2007, p. 442) which, though never fully determinable, should consist at the very least of the absence of the systematic eradication of life's rich assemblages.

Ecotopian Reconfigurations of Terrestrial Relations

In their gestures towards better worlds, many REAs call for fundamental shifts in how we live with others – from systematic annihilation and rigid hyper-separation toward respectful cohabitation (Plumwood 2002):

> I envision a planet of Earthlings, some feathered, some scaled, some hairy, some leafy, but all living in harmony in a world that values beauty over capital, life over property, peace over war. (Captain)

Captain's vision recalls Latour's (2017) emphasis on the need to learn to live well together, through painstaking negotiation with multiplicities of other 'terrestrials' amid the Capitalocene. Modes of being that prioritise respectful coexistence with terrestrial kin are the ideal for many REAs, who call for greater attentiveness to their unique lived experiences, needs and vulnerabilities:

> We really need to reassess our speciesist beliefs, and this idea about what is good and what is proper for animals, and maybe really check for their feelings and for their wellbeing (psychological, etc.). (Forest).

In terms of what nature-culture entanglements in an ecotopian society might look like, the EF! 2017 Gathering pamphlet makes clear that in an ideal world, 'Animals are not *ours* [emphasis added] to "farm", enslave, control, cage, slaughter or accumulate wealth from' (p. 34). Thus, for some REAs, domestic species – seen as living legacies of human attempts at mastery and domination over other terrestrials – would no longer exist:

> within the times of revolution, domesticated or 'domesicrated' animals, I believe that they will have to live with us but we probably should choose to not breed those animals purposely, so those lines of domesicrated animals would either eventually rewild or die out, which I am okay with. (Koala)
>
> we don't get to have pets, we don't get to make animals do things for us; we don't get to, like, train dogs, they just get to do what they want and we just treat them in a very respectful way. (Fox)

Recall in the previous chapter that less exploitative forms of using other animals were still prevalent in ecotopian texts. Owing to their anarchic underpinnings, many REAs exhibit a strong aversion to any form of domination and exploitation, whether in terms of political organisation or in how we relate to terrestrial kin. Particularly problematic here is the use of force and/or coercion in our dealings with others. Most of the REAs interviewed envisioned some variant of a global interconnected system of nature reserves and/or mass rewilding projects for the creation of spaces largely if not wholly devoid of human interference in variations of a 'hands-off' approach (Morton 2010). Thus, one EF! document suggests that a core objective should be to 'recreate lost habitats and reintroduce extirpated predators' (2011). After assisting with the creation of rewilded spaces, the aim would be to let other species and natural processes 'be':

> I'd like to see more wilderness areas, and more areas re-wilded. I'd like to see an end of animal agriculture or at least a big scaling down of it, and part of that sort of dynamic would be taking cows and sheep in this country off the land, letting that land re-wild back, which would then increase biodiversity across the board. There's also this thing like the lynx reintroduction which, they were hopefully going to reintroduce six lynx into the forests of Northumberland, and if they do that the lynx will hunt the deer, and the deer will stop eating all the trees in the forest, and there will be more trees, and more birds, bees, butterflies, smaller mammals, and a massive increase in biodiversity then, kind of like, almost for free. (Squirrel)

> like this half-earth idea, giving half the earth over to nature. You've got places, like where the Chernobyl disaster took place, rewilded incredibly. And making taboo again, like, tracts of land in which humans cannot interfere or exploit, is really vital. (Badger)

> with some places that humans just aren't allowed to go, like, 'We don't get to get to go to this area because that's just where the animals live', and we don't get to bother them all the time ... (Fox)

As a result of their experiences with the disastrous consequences of destructive human (or more precisely, industrial-capitalist-modernity's) interference in ecosystems and spaces of other-than-human habitation, many REAs thus call for spaces exclusively devoted to the flourishing of other-than-human life. Captain similarly urges a radical descaling of human communities, to be interspersed between vast wilderness areas:

> what we need to have is communities no larger than 20,000 people, those communities should be separated by large areas of wilderness, transportation between those communities, we have the technology for it right now, would be through underground systems, you know, between that, to allow as much of the surface of the earth to be controlled by wilderness ...

Key recent developments within biodiversity conservation are helpful to reflect on in relation to the present discussion. Mainstream conservation still bears traces of colonial or 'fortress conservation' (Brockington 2002). This approach within the Western conservation tradition tends to be predicated on dualistic views of humans as separate from 'wild' nature, and has been associated with numerous instances of neocolonial displacements of indigenous and local groups in order to make way for the creation of 'untouched' Nature preserves (Cronon 1996; Büscher et al. 2017;

Dominguez & Luoma 2020). This has been particularly the case in relation to areas designated as 'strict' protected areas (PAs) under The International Union for Conservation and Nature's (IUCN) classification system, which impose strict restrictions on local communities' access to and use of the resource base. Presently, just over 16% of the earth's land surface is set aside as PAs; Vimal et al.'s (2021) assessment of 175,000 global terrestrial PAs found that around 41% of these are under 'strict' control and management. After nearly 200 national signatories failed to meet a single one of the 20 Aichi biodiversity targets adopted at the UN Convention on Biological Diversity (CBD) conference (COP10) in 2010, the UN CBD's new 'post-2020 biodiversity framework' aims to set apart 30% of the earth's land and seas as conservation areas by 2030 in the form of equitably managed and integrated areas (CBD 2022). However, given the largely unchallenged colonial and Anthropocentric legacies noted above, fortress conservation practices may become more frequent. Entreaties for acting quickly on the climate and ecological crisis, as in the recent UNEP (2021) report 'Making Peace with Nature', still often remain fixated on the threats that biodiversity loss pose to *human* wellbeing. The Kunming declaration which resulted from part one of the COP15 negotiations in 2021 boasts the decidedly ecotopian title 'Ecological Civilization: Building a Shared Future for All Life on Earth', as well as calling for 'urgent' and 'transformative' change; yet it too focuses largely on the ways in which biodiversity 'underpin our human and planetary health and well-being, economic growth and sustainable development' (CBD 2021a, pp. 1, 2). The first draft of the UNEP Convention for Biological Diversity's (CBD) post-2020 global biodiversity framework, presented at the second session of the COP15 negotiations in December 2022, seeks to undertake 'transformative actions' in order to reduce threats to biodiversity by 2030 for 'living in harmony with nature' by 2050. Yet this initiative similarly frames 'nature' and other terrestrials as instrumental service providers, wherein the key motivation is to 'preserve and protect nature and its essential services to *people* (emphasis added)' (CBD 2021b).

Once again, the prevailing narrative of endless growth for human-centred development persists, with biodiversity deemed valuable and worth protecting because of its indispensability first and foremost to *human* health, security and prosperity (CBD 2021a, p. 3). This is partly unsurprising, as mainstream conservation tends to be situated within the reformist paradigm[8] of 'sustainable development' which neither seriously challenges Anthropocentrism nor (neo)colonial-capitalist expansionism, and often merely seeks to 'green' the latter. Indeed, this points to a wider (and troubling) trend around the extent to which terms like 'transformative' and 'urgent' are being co-opted by mainstream sustainability discourses at a purely symbolic level, with no such changes meaningfully pursued or put into practice. The World Business Council for Sustainable Development's (WBCSD) latest report 'Vision 2050: Time to Transform' calls for 'radically different mindsets' and 'transformative pathways', and outlines steps for businesses to become more sustainable in response to contemporary ecological crises. It points out that these are not 'utopian ideals' but 'possible and practicable recommendations' in line with existing frameworks such as the Sustainable Development Goals (p. 6). It also generally continues to frame 'Nature' and other species as mere resources whose overarching purpose is to secure our own continuity (Crist 2017; Irving & Helin 2018; Tschakert et al. 2020). Little wonder why a recent meta-analysis (Biermann et al. 2022) of 3,000 scientific studies of the sustainable development goals conducted between 2016 and 2021 concluded that they lack transformative political potential and expressed serious doubts that they could help steer societies 'greater ecological integrity on a planetary scale' (p. 797). Effective and just protection of biodiversity is not merely dependent on 'science' and its facilitation of measurable and enforceable targets (Annette 2022) but, crucially, on profound societal changes for addressing the structural drivers of these losses – i.e. hierarchical and dualistic constructs of humans/other-than-human relations born of Anthropocentric worldviews and socioeconomic systems predicated on the boundless extraction and commodification of earth systems for profits. REAs resoundingly urge against the profound limitations of such self-regarding anthropocentric

---

8 Although the concept was decidedly radical when it first came onto the global scene in the late 1980s. Until then the 'environment' and questions about ecological sustainability hadn't seriously factored in 'development' theory and policy.

motivations. Ask not what good an undrained marsh is to us; ask, what good is it to those who live in and depend on it for their wellbeing?

E. O. Wilson's (2016) decidedly bold 'half-earth' proposal alluded to by some REAs seeks to set apart half of the Earth's land mass for human activity and habitation, leaving the rest for 'nature'. It is the largest proposed system of land reallocation and governance in history (Ellis & Mehrabi 2019). However, a number of potentially serious ethical, political and logistical concerns arise. For instance, what would happen in the 'human' half of the Earth? From whence the resources and personnel to police the human-wildlife divide, and who would have the power to do so? What would happen to the approximately 71% of Earth's terrestrial surface that is currently taken up by agriculture and land for crop cultivation, cities, settlements, and other infrastructure (Ellis & Mehrabi 2019, p. 25)? 'Who will bear the costs and who will reap the rewards of such great global trade-offs' (Ellis & Mehrabi 2019, p. 28)? Of course, in some cases, strict conservation spaces will be necessary for ensuring the continuity of especially vulnerable species. However, in most regions, as Ellis and Mehrabi (2019) point out, we will have to opt for 'shared multifunctional landscapes' wherein different species cohabit and use territories for a range of purposes. In other words, we will have to adopt a spectrum of conservation approaches rather than try to aim for a single silver-bullet solution to what amount to exceedingly complex and contextually variable crises (Rolston 2015). Moreover, radically scaling back industrial factory farming and chemical-intensive commodity-based agriculture and shifting towards, for instance, regenerative farming practices could give other species more room to thrive. Ironically, Oldekop et al.'s (2016) meta-analysis of 165 PAs and factors contributing to their success demonstrates that those with most successful conservation outcomes tend to be those with positive socioeconomic outcomes – i.e. reduced socioeconomic inequalities, the empowerment, active and democratic involvement of local people, and incorporating local or Traditional Ecological Knowledge (TEK). This further highlights the extent to which environmental issues are always also social justice issues (Thomas 2022).

Though motivated by the laudable and vital aim of halting contemporary tides of biological annihilation, 'hands-off' approaches as seen in the

'Half-Earth' proposal (Wilson 2016) and occasionally detected amongst REA prognostic framings overlook the intersecting, structural drivers (i.e. extreme socioeconomic inequality and growth-oriented production and consumption) of climate and biodiversity decline (Holland et al. 2009; Büscher et al. 2017). The 'hands-off' approach rehashes traditional human/nature and human/animal dualisms (Plumwood 2002; Latour 2004), wherein 'wildlife/wilderness' and the 'human/societal' realms are depicted as separate and/or irreconcilable (Cronon 1996). However, as discussed in Chapter 2, if everything is interconnected, then there is no background or foreground, no humans over here and nature 'over yonder', no place for any of us to step 'outside' of intraspecies entanglement (Collard et al. 2015). We drive and fly 'using crushed and liquefied dinosaur bones. You are walking on top of hills and mountains of fossilized animal bits. Most of your house dust is your skin ...' (Morton 2010, p. 269). The goal, then, should be to facilitate more careful integration and respectful cohabitation, wherein the needs of singular earth others are taken into account, an undertaking that I'll explore further in subsequent chapters. The objective of political-ecological movements during the increasingly visible nature-culture-material entanglements of the Capitalocene should be to embrace and learn how to harmoniously navigate such entanglements with our co-terrestrials in our strivings for multispecies justice, not attempt to retreat from them (Latour 2011, p. 21). The supposed zero-sum game between humans and nature leaves us at a dangerous impasse.

Indeed, Shark initially considers the desirability of spaces wholly devoid of human presence, but then reflects on the potential difficulties with such an approach to multispecies relations:

> I actually think that there should definitely be some places on the planet which are not touched by humans, because in this way we can measure our impact ... but how could we not touch an ecosystem? Because there's no such thing as ecosystems if you think about it, in like the broader term, because every ecosystem is connected to somewhere.

Rewilding aims to protect and enhance biodiversity and the dynamism and resilience of ecosystems in an increasingly defaunated world. Though ultimately seeking to minimise human interventions, it also (especially at the outset) features considerable human involvement in the form of species reintroduction, the modification of river flow patterns, promotion of naturalistic fire regimes and the like (Monbiot 2014; Lorimer et al. 2015; Corlett 2016). The prefix 're' connotes 'back to', a notoriously nebulous concept as it is exceptionally difficult to discern what we are returning to since human alterations of natural systems and landscapes have been occurring for thousands of years. This renders conceptualisations of a pre-civilisational, pristine state of unperturbed natural equilibrium more mythical than actual, even though unprecedentedly volatile environmental paroxysms can be detected with the onset of the Capitalocene. Furthermore, natural systems are mosaics in states of constant flux rather than static entities, with shifting equilibriums and population dynamics, and endless cycles of birth, death, and decay. However, crucially, much discourse around rewilding (particularly amongst REAs) is decidedly present and future-oriented. The ultimate endeavour is to replenish, restore, reforest, and recreate species abundance, to 'perpetuate the integrity of evolutionary and ecological processes' (Rolston 2015, p. 34) that are being systematically eroded amidst the Capitalocene. By radically reconfiguring the industrial-capitalist-extractive enterprise (Bar-On et al. 2018), we can create possibilities for the (re)emergence of abundant futures. And, as I've discussed in previous chapters, it is our ethical responsibility to other terrestrials to attempt to mend the damage, to reduce harm and work towards a mutual flourishing, however imperfectly. Rewilding initiatives, if done in the spirit of promoting multispecies flourishing and terrestrial abundance rather than seeking to rehash dualistic conceptions of humans as inhabiting a separate realm to 'Nature', will increasingly become important 'ecotopian' strategies.

As I suggested earlier, many REAs gesture towards the possibilities and desirability of respectful coexistence with other terrestrials, wherein humans learn to 'no longer throw the spider out of the house' (Grasshopper):

> I think that we don't need a strict 'no' to co-living with animals, but we really need to reassess our speciesist beliefs, and this idea about what is good and what id proper for animals, and maybe really check for their feelings and for their wellbeing (psychological, etc.), and then we might find that most of the zoos are not good enough for animals, maybe even all of them. (Forest)

> ... if we build cities where we live in, sort of, very small land-footprints, and we then integrate into the ecology of the surrounding area, if we free that up for wildlife, we live in this very small footprint but we integrate with the wildlife and we try and support it as much as possible, try and increase biodiversity, try and create awareness about what we're doing and try and learn from it and do research, I think that's definitely a positive thing. (Jellyfish)

As in the previous chapter, the above excerpts allude to the intricacies, complexities and challenges that will attend the messy work of learning to co-inhabit our radically altered planet with multiplicities of other beings. Moreover, the excerpt evinces an awareness of and underscores the ultimate instability of rigid binaries and their ultimate failure at grasping the entangled and ever-shifting nature of reality (Latour 2004, 2005). With nowhere to extricate ourselves to, the emphasis instead is on how we can best share our finite terrestrial existence together.

Ecotopian Politics

Extinction Rebellion has notably called for the establishment of citizens' assemblies modelled on models of direct-democratic political participation (Simpson 2021). Herein, citizens from an array of cultural, gender, experiential, etc. backgrounds are randomly assembled (by a team of impartial coordinators) at the local or city level for listening, deliberating, and deciding on matters of key social significance (Extinction Rebellion Citizens' Assemblies Working Group 2019, p. 7). Although XR's invocation to 'go beyond politics' constitutes a serious flaw in their approach, as the business of instituting fundamental societal transformations for mitigating the climate crisis is an inherently political undertaking (Alberro 2020). Indeed, as EF! UK succinctly puts it in a recent post titled 'To our friends from XR: Rebel, rebel!' (2022):

> Politics isn't about politicians and parliament, it's about people trying to create the society they believe in ... taking action on climate change is fundamentally political ... De-politicisng climate change not only ignores capitalism and other root causes, it also means XR will continue to be dominated by white middle class 'environmentalists'.

Going beyond institutional or parliamentary modes of political participation, on the other hand, which XR do to a degree, is indeed necessary. Other REAs and ecotopian texts similarly reject institutional political modalities characteristic of Western liberal democracies (Plumwood 1995) and emphasise the desirability of direct-participatory modes of socio-political organisation, citing eco-anarchist 'utopian' enclave in Rojava (van Klink 2021) as an example:

> the best example we've got at the moment is Rojava [?] in the Middle East in West Syria, West Kurdistan, and it's an autonomous community of five million people that are organized along this sort of horizontal line, so there's direct-democracy and organizing by consensus, and also interesting that they're kind of like quite ecocentric, and also they have these sort of structures where you have these communes, and they're kind of small, and then you go up to assembly, and then assemblies might be like an environmental assembly or one that deals with healthcare, one that deals with education, and interestingly they have, like, mixed ones and female ones, and the female ones get the veto. And I think that's quite an interesting model in society. (Squirrel)

> The concept of property is rejected, with working class appropriation encouraged – via land squats, occupations and more. In the long-term process of social change, there is a commitment to the recollectivisation of land 'ownership' and the creation of space for diverse patterns to be explored, suited to each community and bioregion. (EF! 2017, p. 33)

As noted previously, REAs engage in thoroughly contentious and extra-parliamentary modes of political participation, as indeed must all liberation struggles mobilising against unjust systems and legal frameworks. This is very much reflected in their visions of ideal political formations. Direct-democratic modes of political participation, as well as more communal family and group relations, serve to undermine traditional (largely Western liberal democratic) and rigid distinctions between the 'public' and the private' sphere (Plumwood 2007). Such responsive modes of democratic governance and participation, which emphasise equal communication and decentralised decision-making rather than entrenching power and influence in the hands of an elite minority (Plumwood 1995, p. 137), are regarded as most congruent with social and ecological wellbeing. Although, such modes of political participation can also be taxing in terms of the extent of involvement required. A radical reorientation of work, organisations, enterprises, and institutions more generally – from profit-oriented production and conspicuous consumption towards provisions for multispecies wellbeing – are deemed essential for the establishment of more 'ecotopian' ways of being (Marcuse 2013).

REAs' Terrestrial Ecotopianism

> Can we relearn to live in the time of the end without tipping thereby into utopia, the utopia that has beamed us into the beyond, as well as the one that has caused us to lose the here below?
>
> – Latour (2017, p. 285)

Western modernity has served as an apocalypse in the sense of putting an end, often violently, to other species, civilisations and onto-epistemological modalities (Latour 2017, p. 205). This is especially so amidst the Capitalocene, which is seeing the end of numerous worlds as climate and ecological catastrophe continue to diminish life's rich tapestry. Crucially, to live in the 'end times' means taking responsibility for the worlds we have unjustly imposed an end on (Latour 2017, p. 285), seeking ways to co-construct more ethical and inclusive relations with our fellow terrestrials. Hence the need for terrestrial modes of utopianism as opposed to transcendental modalities that – predicated on Western-technocratic notions of a de-animated Earth – project their visions of the good into 'the beyond'. By contrast, the permeation of crisis and uncertainty into the present has compelled REAs to 'kick' ever more vociferously in the 'here and now' (Garforth 2018, p. 158). As the discussion above has denoted, their oppositional mode of activism is reminiscent of the 'critical utopias' (Moylan 1986) of the 1960s and 1970s. Especially palpable amidst contemporary REAs mobilisations and ecotopian sensibilities is a largely post-left anarchist aversion to majoritarian politics and an uncritical embrace of industrial society, modernist utopian separations of the present and future, deferrals of 'the better' to the latter, and grand blueprint designs. Rather, they avow the urgent and immanent realisation of more just worlds through concrete actions and relations in the present

(Robinson & Tornmey 2009, p. 158). However, such groups also embrace characteristics of the earlier left such as valuing equality, autonomy, and critiquing capitalism and alienation (Robinson & Tornmey 2009, p. 159). As Robinson and Tornmey (2009) aptly summarise it, the utopianism of post-left anarchist movements manages to be 'at once idealistic and radical but which also calls upon us to act in the here and-now' (p. 157). I posit that it is the pervasiveness of ecological collapse and the loss of cherished terrestrial kin that especially adds a unique impetus to the 'nowness' of REAs' sensibilities and mobilisations by fundamentally altering their relations to hope and futurity.

What remains of interest is how amongst many REAs, collapse and loss engender not passive acquiescence or escapism but rather radical critique of – and fervent, redoubled resistance to – the Now in the form of direct-action tactics, anti-capitalist narratives, and ecotopian subjectivities. REAs' post-anthropocentric worldviews and kinship bonds with Earth others, though occasionally failing to transgress dualistic and hierarchical constructs of human relations with other species, are central here (Sargisson 2002; Pepper 2007). Affinity for the other-than-human world appears to have been internalised to the extent that it fuels their fervent mobilisations for stemming the further loss of Earth kin. REAs are simultaneously situated within the current eco-apocalypse and mobilised by an intense passion to resist it (Latour 2017; Alberro 2021), suggesting that even apocalyptic conditions harbour seeds of hope (Jameson 2005 199). However, hope of the kind discussed presently and alluded to by Solnit (2016) in the introductory chapter – the kind that shoves you out the door during an emergency, and that asserts that a better world, though never guaranteed, is always possible. In citing a profound hopelessness, REAs eschew not hope as such but hope in an abstract form, located in distant times and places (Bloch 1986), the disembodied hope of 2030 and 2050 climate mitigation targets. Hence, their repeated emphases on the need to live and embody better alternatives within the Now. REAs' 'critical modalities of hope' (Duggan & Muñoz 2009) feature a dialectical interplay between a concrete hope (Bloch 1986) and 'negative energetic hopelessness', wherein shared feelings such as cynicism, despair, and grief over widespread loss of cherished Earth others are 'critically redeployed' (Duggan & Muñoz 2009, p. 278;

Pike 2016) toward a fervent refusal of the Now and its myriad injustices. REAs thus embody a 'terrestrial' (Latour 2017) modality of utopianism that is decidedly immanent, rejecting deferral to distant times and places, and firmly rooted in terra and in relation with other terrestrial kin. Such modes seek to dismantle the 'Now' as presently constituted and enact the unrealised potentialities of an animated materiality in constant flux.

REA relations to the heterogenous 'other' – in the form of other-than-human life and the 'Not-Yet' – that to varying degrees problematise (though don't always succeed in deconstructing) hierarchies and rigid boundaries strongly echo Derridean deconstruction (2016). Deconstruction, though lacking a single or precise definition, crucially seeks to expose, reverse and displace hierarchies and binary opposites so as to 'pull them apart' (Derrida 1995b, 2016). Largely focused on embodying 'militant and interminable political critique' in the 'Now', REAs proceed with considerable caution towards notions of 'the better' and the 'future-to-come'. They often alternate between a messianic openness to the unforeseeable 'future-to-come' (Derrida 1994) through their aversions to closure while still expressing a desire for *some* 'content' – i.e. more egalitarian modes of co-terrestrial relationality devoid of the systematic eradication of life's rich assemblages. Thus, REA aversions to closure (Gordon in Davis & Kinna 2009) stem not only from their post-anarchic underpinnings but, crucially, from their avowals of an irreducibly heterogenous future which, like the 'other' and notions of a complete society or fulfilled history, can never be fully anticipated by finite and historically situated subjects (Derrida 1994). Theirs is a form of 'undetermined messianic hope' (Derrida 1994, p. 35), a messiahless messianism that doesn't anticipate any single agent of history (Latour et al. 2018) but mobilises in answer to the ethical pull of threatened earth 'others' (discussed below). In other words, messianism in the Levinasian sense of ethical openness to the singular 'other', an inextinguishable hope in the enduring possibility of a better to-come (Derrida 1994, 2005b). The latter is, again, rendered possible because, like our heterogenous terrestrial kin, futurity forever resides beyond the bounds of full approximation (Lazarus 1999). Moreover, REAs' complex relationship to utopianism's compensatory dimensions is very much a reflection of the 'end times' of Capitalocene socio-ecological breakdown. These times are thoroughly 'out of joint' (Derrida 1995b), plagued by proliferating uncertainties and entanglements. The intrusion of ecological collapse into the present have shattered traditional conceptions of linear historical and temporal movement. Thus, REA ecotopian modalities effect an absorption of the utopian horizon into the precarious present, where 'ecotopia' is embodied via imperfect everyday practices rather than situated in a post-revolutionary future scenario (Gordon in Davis & Kinna 2009, p. 261).

At times, the eco-apocalypticism that surfaces in their imaginaries serves to extinguish the undetermined messianic promise of the future-to-come by proclaiming the inevitability of wholesale eco-catastrophe, a move that effects a closure of the heterogenous 'Not-Yet' and denies the recalcitrant nature of reality. However, as we've seen in this chapter, in their strivings for 'justice' – for those presently living, those 'who are no longer', and 'those who are *not yet present and living*' (Derrida 1994, p. xiv) – they assign *some* tentative content to the 'better' and the 'Not-Yet'. They are unequivocal about their concrete desires to avert a 'Not-Yet' devoid of life in its multifarious manifestations, which would be dystopia par excellence. As Friberg (2022) denotes in her analysis of recent climate mobilisations such as Extinction Rebellion as instantiations of a 'disruptive utopianism', such groups are driven by a present catastrophism to 'create a gap between the space of experience and the horizon of expectation that makes it possible to transform the future by adjusting things in the present' (p. 8). Though we should remain open to the surprise of new and ever-shifting alliances, deliberations surrounding how to live well together require some degree of closure, otherwise we would never succeed in knowing what a better world *could* look like (Latour 2004, p. 111). In other words, despite the limitations and violent exclusions effected by boundary delineations, the exigencies of Capitalocene crises render it existentially as well as ethically crucial for *some degree* of determinateness around desired alternatives without closing the door entirely on heterogeneity. If not, we risk a 'pathological pluralism' that fails to direct social change towards the better (Levitas 2000, p. 40; 2003, p. 148), which we can little afford amidst extant climate and biological breakdown. REAs problematise rigid boundaries between the present and to-come, simultaneously embodying, desiring and gesturing towards worlds devoid of exploitative and reductive relations with our terrestrial kin. REA movements live at the intersection of dread and hope (Kirksey et al. 2013), as pockets of radical ecotopian potential that both resist the depredations of the Capitalocene whilst gesturing towards worlds for intraspecies flourishing.

CHAPTER 6

For the Love of Kin: Indigenous Green Futurisms

> Utopias are what we have to build, and build now, in order to find some type of sanctuary in which we and all others can live – there is no plan or planet B for us to turn to.
>
> – Whitehead 2020, p. 12

No Plan(et) B, indeed. Two-spirit novelist and member of the Peguis First Nation Joshua Whitehead reminds us in their introduction to *Love After the End* (2020), an anthology of two-spirit and indigiqueer speculative fiction, that indigenous peoples have already survived an apocalypse – that imposed by Western, colonial-capitalist modernity. The (re)intrusion of socio-ecological collapse into the present seen within the Capitalocene as alluded to by REAs in the previous chapter is something that many local and indigenous peoples have been reckoning with for centuries. Indigenous peoples have long lived through the violent legacies of colonialism in the form ecosystem collapse, species loss and forced relocations resulting in the severing of honoured kinship relations, economic and cultural disintegration (Whyte 2018). The sheer scale of this violent decimation of worlds is difficult to overestimate – 56 million people, accounting for 90% of the pre-Columbian native population across the Americas, had been eradicated by 1600. Known as 'The Great Dying', this is one of the largest known mass human mortality events relative to global population, second only to the Second World War, an estimated 80 million people, equivalent to 3% of the world's population at the time (Koch et al. 2019). The loss was so extensive that it resulted in a dip in global atmospheric CO_2 concentrations.[1] As such, Mykaela Saunders, whose

---

[1] Similarly, the often-cited decimation of the American buffalo across North America was not the result of native over-exploitation of the species. Estimates of precontact populations of the North American buffalo range from 25 to 30 million. As Sioux

fantastic anthology of Aboriginal and Torres Strait Islander speculative fiction *This All Come Back Now* (2022) is discussed later in this chapter, poignantly notes that many common themes and tropes in speculative fiction around apocalypticism are just 'stone-cold reality for us' (p. 9). She observes how, whilst *Mad Max* may be one of the best-known Australian cli-fi stories, 'for our people, who have seen unfathomable ecocide enacted hand in globe with our own attempted genocide, all stories that take place in unceded lands post-1788 are climate fictions' (p. 9). The European imperial project at the heart of such catastrophes was in part facilitated by the antagonistic racial, cultural and gendered 'othering' of 'non-European' cultures, and a will to control and manipulate such otherness for socioeconomic, cultural and political supremacy (Said 1995). Indeed, the very Western/non-Western binary serves a similar function of violent erasure of singularity to the human/nonhuman binary discussed in Chapter 2 (Plumwood 2002; Dutton 2010).

Despite such destructive legacies, other ways of knowing, living and relating to terra persist. Thus, this chapter aims to explore the potentiality laden within the sensibilities, strivings and select literary works of indigenous peoples who, both historically and presently, have long embodied radical alternatives to Western modernity. However, in the discussions that follow I do not wish to perpetuate a monolithic image of all indigenous peoples as living in perfect harmony with terra. This form of 'ecological exoticism' (Ferdinand in Pellizzoni et al. 2022, p. 50) fosters 'the exclusion of the racialised other as a companion on equal footing with whom a common world is imagined' (p. 50). The incredible diversity of indigenous cultures historically and presently belies any such attempts at homogenisation or reductionism. For instance, McWethy et al.'s (2009) examination of pollen

---

scholar Nick Estes recounts (2020, p. 78), white settler-colonial attempts to appropriate native land occasionally took the form of the mass extermination of buffalo- a key food source and close relative of the Plains Nations- in order to weaken Indigenous resistance to colonial expansion. The remaining 10–15 million animals were wiped out within decades. Accounts like these serve as a stark reminder of the violent socio-ecological legacies of colonialism, and further belie uncritical designations of all of humanity as equally at fault for contemporary biodiversity and ecological breakdown.

and charcoal records from New Zealand denote 'striking evidence' for an association between initial Polynesian (Māori) arrival around c. ad 1280 and widespread burning and loss of native forests on the island (p. 884). Moreover, many indigenous peoples whose traditional cultural and social practices have been profoundly altered and distorted by the damaging legacies of colonialism and capitalism, engage in environmentally destructive extractive projects. Bolivia's former (and first) indigenous president Evo Morales, though many of his policies benefited Bolivia's majority indigenous population, also doubled down on extractive development models (Postero 2017). Moreover, as Braidotti (2013) observes, other-than-human animals since antiquity have often served as a sort of 'zoo-proletariat' at the bottom of the species hierarchy, variedly exploited for labour, their flesh, for amusement and our benefit, and this pattern is not exclusive to Western cultures (p. 70). Though, as we saw in Chapter 2, the tradition of antagonistically 'othering' other-than-human terrestrials and reducing them to the status of exploitable objects with little more than instrumental value has been especially prominent in the West. What's more, it's important to recall that utopianism is not about perfection or silver-bullet solutions – it is about imagining and striving towards 'the better', a process which is inherently pluralistic, always-ongoing and inevitably laden with challenges and inconsistencies.

As such, I only wish to shed light on a few examples to demonstrate that, on the whole, many indigenous cosmologies, and ethico-political modalities – which tend to be guided by core principles of care, respect, reciprocity and attentiveness – gesture towards more ethical and generative modes of being within and beyond the Capitalocene. Many of these societies tend not to engage in the sort of systematic waste and destruction of the natural world generally seen within Western capitalist societies and, crucially, tend to relate to other terrestrials as agentic kin rather than external objects or commodities. The lands of indigenous peoples the world over remain battlegrounds with multifarious attempts to halt the further advancement and intensification of extractive industries (Kröger 2021). From the Movement for the Survival of the Ogoni People's (MOSOP) successful expulsion of Shell from the Ogoni region in 1993 after decades of fighting for territorial and resource sovereignty (Temper 2019), to the multi-tribe alliances in the Amazon basin presently resisting the encroachment of illegal miners on their lands. Indigenous transgressions against the deficient present also surface via their alternate ecological sensibilities, diverse gender identities and in the form of organisations such as UTOPIA (United Territories of Pacific Islanders Alliance), a grassroots social justice organisation based in Washington, US led by LGBTQIA+ Pacific islanders. As I'll discuss in this chapter, like REAs, many exhibit a terrestrial ecotopianism which seeks to build places of refuge in the here and now, in respectful relation with our myriad terrestrial kin. The latter furthermore tend to be framed not as mindless, soulless automatons but as agentic and purposive persons. In the discussion that follows I explore some examples of indigenous ethico-political and onto-epistemological modalities, as well as literary works in order to shed further light on the potentialities of indigenous ecotopianisms. In doing so, I proceed from a stance of solidarity and humble curiosity, with an awareness of the limitations of my own positionality as a researcher situated in the 'West'. I also acknowledge that all knowledge is situated and therefore ultimately limited, and proceed with care when attempting to write about experiences that are not my own. Here, as with all investigative endeavours, I recall science fiction writer and editor of the wonderful anthology *Invisible Planets: 13 Visions of the Future from China* (2016) Ken Liu's sobering reflection: when attempting to interpret literature from one's own culture, let alone another, it is wise to proceed from the acknowledgement that we don't know much, that we need to learn a lot more, and that there are no simple answers (p. 15).

Segue into the Pluriverse

> There are multiple pathways, multiple worlds, multiple ontologies – a pluriverse.
> – Larsen and Johnson (2017, p. 28)

Mark Fisher's poignant work, *Capitalist Realism: Is There No Alternative?* (2009) features a timely and erudite expose of our dominant socio-economic system that 'occupies the horizons of the thinkable' (p. 8) by masquerading itself as the only viable option, hence our collective difficulties in imagining anything outside of its 'bleak' borders (Klein 2019). For Italian Marxist philosopher Franco Berardi, 'power' as manifest in and deployed by hegemonic systems like capitalism, colonialism and anthropocentrism, is an essentially anti-utopian force involving the 'selection and enforcement of *one* possibility among many, whilst simultaneously excluding (and invisibilising) of many other possibilities' from a given space of actualisation (2017, p. 2, 6). Yet, capitalism is a relatively novel system in historical terms – other socioeconomic forms preceded, exist alongside of, and in time will come after it. Pluriversal political theories, encapsulated by the excerpt above, remind us that 'reality' and possibility are pluralistic and dynamic; this is a world of many worlds, actual and possible. The very existence of other-than-Western worlds and the potentiality they serve as a testament to constitutes a powerful utopian challenge to the capitalist-realist conviction that there are no alternatives. Unearthing and exploring such ways of being is part of the vital decolonial process of 'freeing the world from universal claims of Western epistemology, dressed as superior knowledge' (Reiter 2020, p. 104). A rich body of anthropological research[2] shows that, in stark contrast to Western anthropocentrism, many Indigenous peoples, from Brazilian Amazonians to the Alaskan Koyukon (Watson & Huntington 2008) and Australian Aboriginals (Bird Rose 1999; Suchet-Pearson et al. 2013), see the wider terrestrial world as 'animated, agential, knowing, feeling ...' (Plumwood 2002; Anderson 2010; Yazzie 2018; Celermajer et al. 2020, p. 6; Escobar 2020; McGregor et al. 2020). Many have thoroughly *relational ontologies* wherein living and ancestral terrestrial persons are seen as kin who exist in inextricable entanglement with one another.

In Ojibwe ontology, the concept of personhood doesn't begin with the human self and extend outward but rather begins with personhood as a broader multispecies category within which humans form merely one subgroup (Harvey 2005; Rose 2013). Different persons are seen as expressing

---

[2] With a caveat that anthropology as a discipline and approach has also often been steeped in colonial tendencies. The works of Deborah Rose Bird and Thom van Dooren are some notable exceptions to this.

different degrees of power or agency, though this does not equate to a hierarchical value ordering. For the Oglala, one of the Lakota tribes, personhood applies to 'anything that has being' and therefore is 'capable of relating'; so, there are human persons, plant persons, and even stone persons, all of whom have both ontological and moral significance (Detwiler 1992, p. 239). Persons are those we interact with, with 'varying degrees of reciprocity', and who have their own perspectives (Harvey 2005). Indeed, for the Oglala, plant and other-than-human animals are regarded as more powerful than humans because they are purportedly better at relating harmoniously to other kin and the world around them (Hall 2011, p. 110). The lack of humility prevalent throughout many contemporary Western societies, evident in a failure to harmonise the needs of different kinds of persons, on the other hand, is regarded as a destructive mode of personhood at the core of ecological decline (Hall 2011, p. 111). Many precolonial African traditions were similarly suffused with a communal rather than an individualistic ethos, ecological awareness, and tended to regard other species, plants, and the like as persons rather than inert objects (Appiah-Opoku 2007; Mwangi 2019, p. 29). Thus, the African philosophy of *ubuntu* evinces a posthuman ethic grounded in concern for the wellbeing and interests of human and other-than-human peoples (Mwangi 2019). Based on the Xhosa proverb 'Umuntu ngumuntu ngabantu' (I am because you are, you are because we are, as opposed to the Western 'I am because you *are not*'), *ubuntu* celebrates the profound relationality that characterises existence, a relationality that spans between currently living persons as well as ancestors and those yet to come. As discussed in the close of Chapter 2, we cannot be without the multiplicity of others who make us possible. An important political implication of the *ubuntu* principle is that societies ought to be run by and for, and nurture all who constitute a given collective.

Activist and academic Nick Estes, citizen of the Lower Brule Sioux tribe in South Dakota, US, notes that many Lakotayapi nouns such as 'Mni Sose' (roiling water) indicate 'not merely static, inanimate form, but also *action* [emphasis added]. In this landscape, water is animated and has agency; it streams as liquid, forms clouds as gas, and even moves earth as solid ice – because it is alive and gives life' (2019, p. 9). Moreover, ethics based on 'reciprocity, respect, sustainability and spiritual-material interconnection' echo throughout Indigenous cultures across the globe (Celermajer et al. 2020, p. 7). Unlike Western epistemological strivings for objective and universally applicable knowledge, Native American epistemology, as denoted by the Native American Tutchone term *tli an oh* – meaning 'responsibly true' (a 'responsible truth') – is *ethically oriented*. Responsible truths are those that are 'good for you *and the community*' and 'ring true for everybody's well-being' (Hester & Cheney 2001, p. 320). Moreover, indigenous epistemology tends to be experiential, wherein knowledge flows not from occupying a detached, transcendental subject position but from intimate involvement in – and keen awareness of – the world. More than this, in order to truly understand and know 'realities', it's essential to have respect for the dynamic and multiple kinship relations that constitute it. Thus, for the Navajo – a Native American tribe of the Southwestern US – the concept of *Hózhó* is central, which denotes 'right relations of the world', including human & other beings, 'who are of the world as its storied and dynamic substance, not in the world as a container' (Haraway 2016, p. 91). In other words, 'world' is not something that is occupied, it's something that is made and sustained by myriad agencies with shared ancestries in perpetual cycles of birth, death, decay and resource exchange. Behaving ethically means being a good ancestor and relative, taking care to anticipate how our ancestors and future generations might interpret our actions (Whyte 2018, p. 230). Crucially, the dead in many indigenous belief systems are never wholly departed; ancestors and relatives who've long since passed continue to underlie the structure of the living present as spirits who remain entangled in the lives of the living, as evinced by a local poem recited by one of the participants in Seth Appiah-Opoku's investigation (2007) of rural Ghanaian indigenous belief systems:

> The dead are not dead. They are in the trees that rustle. The dead are not dead. They are in the wood that groans. The dead are not dead ... they are in the water that runs. The dead are not dead. (p. 92)

The origin stories of Aboriginal Australians and other indigenous peoples often feature 'Dreaming beings' or 'spirit ancestors' who emerged from terra – rather than from 'above'. These beings give local landscapes their present configuration and represent the shared ancestral origins that underpin existing kinship relations between persons (Hall 2011). Other animal and plant relatives frequently appear as key agents in indigenous cosmological systems, as in the Kanaka Maoli (indigenous Hawaiians) creation story 'Kumulipo', wherein sperm whales played a significant role during the early emergence of animal and plant relatives far before the appearance of humans (Puniwai 2020). Many indigenous cultures don't merely hold animist views wherein the world is perceived to be full of human and other-than-human persons (Ingold 2006; Hall 2011, p. 100). Rather, they *are alive* to the world, embodying a 'heightened sensitivity and responsiveness, in perception and action, to an environment that is always in flux' (Ingold 2006, p. 10). Moreover, agency is no mere attribute assigned to materiality, but is ontologically prior to their separation and is diffused across the entirety of the terrestrial world, such that even the winds themselves are seen as persons with agentic abilities of their own (Ingold 2006, p. 16). Such sensibilities are reflected in the languages of many indigenous peoples, as alluded to above. Robin Wall Kimmerer, environmental biologist and member of the Potawatomi nation, poignantly reflects on the power of language as a mediator of how humans perceive and relate to other terrestrials in *The Democracy of Species* (2021). She denotes how English, a noun-based language which divides things between animate (this category usually being reserved for humans) and inanimate, differs so starkly from Potawatomi along with many other indigenous languages. In Potawatomi, a bay, a river, a moose is not a noun, denoting an inanimate entity, but a verb – to *be*. Water is a living entity who is – at one point *choosing* to be a river, at another rain. Water isn't assigned the capacity for agency, it *is* agentic. The language, in other words, is suffused with possible verbs which refer to a world that is thoroughly *alive* (p. 15). Most importantly, mountains, rivers, and other terrestrial beings are never referred to as 'it', which 'robs a person of selfhood and kinship' (Kimmerer 2021, p. 16). When we categorise a tree as an *it*, we erect barriers between us and absolve ourselves of our moral responsibilities to them, effectively 'opening the door to exploitation' (Kimmerer 2021, p. 20). The status of 'it' reduces a living land 'to mere resources. If a maple is an *it*, we can take up the chain saw. If a maple is a *her*, we think twice' (p. 20). An 'it' cannot be a 'fellow', but a 'him', 'her' or 'they' can. As such, the same words are used for humans as well as other terrestrials, all of whom are considered relatives. Kimmerer therefore calls for a shift towards a 'language of animacy' which respects other terrestrials as sovereign *peoples* and which might thereby open the door to more egalitarian and democratic relations.

Suchet-Pearson et al.'s (2013) fascinating co-investigations with the Aboriginal Yolŋu of North-eastern Arnhem Land, Northern Australia, warrant greater discussion as a promising example of an ecotopian ethics and politics for the precarious present. Similarly to other examples discussed thus far, Yolŋu ontology is one of 'co-becoming', wherein humans feature as *one small part* of a broader cosmos populated by diverse beings and ways of being, including other animals, winds, dirt, and sunsets (Suchet-Pearson et al. 2013, p. 185). *Wetj* (sharing), a foundational Yolŋu concept based on an ethic of *responsibility* and rooted in the specificities of particular times and places, is exhibited by the following excerpt which is worth quoting at length:

> The miyapunu (sea turtles) are special animals to us. We have gathered them in this way for as long as this world has existed. We know how to make sure that we don't take too much. We respect those miyapunu and their own lives. We see that their wellbeing and our wellbeing are connected. That is part of the great pattern of kinship ... We care for them and they for us. We sustain them, and they sustain us ... (Suchet-Pearson et al. 2013, p. 191)

The Yolŋu, as do many other indigenous groups, adopt a much more expansive notion of kinship than that typically seen in Western cultures, wherein 'kinship' has traditionally included the family as the primary relational unit (Chatzidakis et al. 2020) and rarely extends beyond the human except for a few companion species. The Yolŋu, on the other hand, evince ethico-political modalities marked by acute attention to, care, respect and responsibility for the myriad and agential terrestrial kin with whom they intra-are in ever-shifting entanglements. This orientation of respect for – and responsibility to – other terrestrials helps ensure that they don't 'take too much', thereby diminishing the capacity for others to flourish and sustain themselves in the long run. Another promising example is offered by the Maya of the Yucatan, who hold that a properly socialised person is one who does not take too much, and wherein over-exploitation and avarice are aberrations considered as prime offences rather than competitive advantages. Anthropologist E. N. Anderson, who has spent some time with the Maya and other indigenous peoples throughout his research, observes: 'I have seen Maya, including my co-author Felix Medina Tzuc, carefully move bugs and ants out of the way of foot traffic' (2010, p. 37). This is the sort of orientation needed in order to respond effectively to the exigencies of the present, wherein 'we' – especially the global plutocratic elite, inhabitants of the Global North, and others in relative positions of power – have a responsibility to undo the oppressive and exploitative 'cultural dominants' of capitalism and humanism (Oppermann 2018, p. 10) that have stolen the present and future from so many earthly collectives. Such changes require shifts in attitudes and worldviews, in who, how and what we value, ethical (re)orientations wherein we develop a new 'hypersensitivity' or 'active attentiveness' (Rose 2013) to the living world around us, and, as discussed below, how we conceptualise the collective 'we'. And promising alternatives already exist in many of our indigenous counterparts' expansive cosmologies, wherein the collective 'we' includes all manner of agents – from rocks and storms to birds and humans.

The Māori are an indigenous Polynesian people of mainland New Zealand. Traditional Māori society was based on kinship and further underpinned by an 'economy of affection' based on gift exchange and reciprocity (Spiller et al. 2011). Māori philosophy – *whakapapa* – stresses (inter) connectedness with spiritual, living and ancestral kin – i.e. plant, animal, waters and lands – understood through their shared genealogical origins and reciprocal relationships (Roberts 2013). Mountains and rivers are not 'impersonal subjects' but living, conscious beings, ancestors with whom humans are inextricably linked and to whom they are responsible (Te Aho 2019). Establishing 'right relations' with the world requires embedding such concepts as *mauri* (life force), *tapu* (potential to be) and *mana* (respectworthiness) into one's interacts with others, and within wider systems of social and political organisation (Watene 2016; Celermajer et al. 2020, p. 11). Māori scholar Linda Te Aho (2019) notes that prior to European contact, tangata whenua (Māori for 'people of the land', in reference to the relationship that tribal groups in the region had developed and maintained over centuries with their ancestral lands and waters) observed a system of laws and values that defined the ethical responsibilities of users of natural resources to preserve the latter for future generations (Durie 2014). All of this was upended following Britain's colonisation of New Zealand, and their establishment of a legal system rooted in Anthropocentric worldviews, and pervaded by alien property relations (specifically individual ownership). Owning and managing 'natural resources' such as rivers for economic gain was entirely at odds with the Māori ethic of protecting the environment for its own sake, as well as for present and future generations to use and enjoy (or kaitiakitanga, the root word 'tiaki' meaning to care for, foster and nourish) (Te Aho 2019, p. 1616). As Te Aho denotes, 'colonization eradicated traditional tribal structures and ways of life, including possession and control of lands and waters' (2019, p. 1618). Competing values and clashes between 'managing' rivers and other natural entities for economic gain versus for cultural, social and ecological integrity have continued. However, in 2017, the Māori succeeded in securing legal personhood for the Whanganui river with the Te Aawa Tupua Act. Te Aho observes that the new act 'gives emphasis to the profound relationships between the river and its people and is intended to provide an opportunity for more effective recognition of the *rights and interests of the river itself* [emphasis added]' (2019, p. 1618). The statement, 'I am the river and the river is me' gives a sense of this new approach to freshwater management. Many indigenous peoples the world over have lived in far less exploitative ways with terra and our terrestrial kin for thousands of years. This legacy has been further corroborated by landmark new research which found that biodiversity, or species richness, is often higher in Indigenous-managed lands, from Canada to Brazil, than in traditional protected areas (PAs) (Schuster et al. 2019).

The Honourable Harvest

> You must care for them [sheep]. You burn incense from ceremony plants for the sheep. The sheep have their own ceremony ... this is how sheep is life. Sheep hide is a bed to sleep on. The way the sheep have sustained us, the thought of herding sheep

is meant to teach you about life ... (Oscar Whitehair, member of the Forest Lake Chapter, Navajo Nation, cited in Benally & Denetdale 2011, p. 69)

Respect, care and attentiveness for others and the earth who sustain us are key components of the indigenous set of ethical principles for guiding 'right relations' with terrestrial kin, and exchanges of 'life for life', otherwise known as the 'Honourable Harvest' (Kimmerer 2013, p. 33). As heterotrophs who cannot produce our own life-giving nutrients or energy, us animals must take life in order to survive. Recalling Plumwood (2003) in Chapter 4, this is a fundamental aspect of our entangled lives with others and is not in itself problematic; to suggest so is to extricate and situate oneself aloof and aloft from the ecological networks within which we're all situated and which we co-constitute. No organism can be without influencing its wider environment and others in some way. But *how* we use others in the business of living matters a great deal. As Le Guin puts it in her epic sci-fi series *Earthsea* (1971–2003) which poignantly explores the limits of human understanding and power over earth others, and avows our inextricable entanglement with them, the issue is the 'unmeasured desire for life', which translates into a craving for 'power over life – endless wealth, unassailable safety, immortality – then desire becomes greed' (1971, p. 333). Present modes of consumption and production within the Capitalocene – especially on behalf of the richest denizens – perpetuate a 'Dishonourable Harvest' based on unchecked desires which can only be satisfied via the theft of ever greater portions of the earth's resources from other living and future generations. This parasitic relationship is dangerously one-sided. Instead, the 'Honourable Harvest' urges that we harvest other plants and animals that we need for sustenance with care and respect for them as well as others, ensuring that we take only that which is given, never take too much, and always leave enough (at least half) for others. Knowing, for instance, that orcas inhabiting the seas of the Pacific Northwest depend on Chinook salmon as a key source of food, we'd take care not to overexploit the salmon in the region, not only for the salmon's sake, but for the sake of the orcas and any other dependents as well. Yet thus far industrial fishing practices have proceeded along the assumption that the oceans' fish, so callously reduced to mere 'stocks', are there for our use alone. Other principles include building a keen awareness of – and sustaining – those who support and care for us, harvesting in ways that minimise harm and waste, and exercising gratitude for that which is given to us (Kimmerer 2013, p. 41).

Indigenous land and natural resource 'management' practices offer further examples of the 'honourable harvest' in action, and demonstrate the important implications of ethical sensibilities rooted in care and respect. In *The Pursuit of Ecotopia: Lessons from Indigenous and Traditional Societies for the Human Ecology of our Modern World* (2010), E. N. Anderson discusses examples of what he terms 'moral conservation' practised by native peoples of the North-western United States, specifically the concept of 'culturally modified trees' (CMTs). Many Northwest native peoples still remove the bark from trees for a variety of purposes such as food, fibre, and as a key component in ceremonial practices. However, typically only one narrow strip of bark is removed from each tree, no matter how rare or far apart, so that each tree can recover and be used time and again by humans and other terrestrials alike. Prior to removing the bark the people recite a 'short but deeply felt prayer' and thank the tree for providing it (Anderson 2010, p. 35), alluding to the important role that gratitude plays in indigenous ethical modalities as a way of honouring relations. In a similar vein, a common Aboriginal practice when harvesting yams is to leave part of the plant in the ground and cover it back up with soil so that it will regrow and continue to flourish and nourish others (Hall 2011, p. 112). This is the aforementioned ethic of respect, care and reciprocity in practice; even where management and harvesting of environmental resources is more intensive, this ethical orientation tends to discourage wanton waste and systematic exploitation. Because there is a profound respect for other terrestrials as agentic kin rather than mere resources, great care is taken to ensure that the plants, other animals and resources that humans depend on aren't squandered or annihilated beyond recovery. Evoking some of the key themes denoted in Starhawk's *The Fifth Sacred Thing* in Chapter 4, for many peoples, including Mongols and Tuvans, forests and mountains are sacred, harbouring powerful spirits. As such they tend to be carefully preserved as sacred entities beyond wanton exploitation and private ownership (Anderson 2010, p. 36).

Jainism is an ancient Indian religion which attributes souls to all living things and even some 'non-living' entities like air and water', and emphasises the reduction of harm and violence to these entities as far as possible in daily life (Dundas 2003). However, an important point of departure between Jainism and the animist orientations of many other indigenous cultures is that in the latter a recognition of other species and plants as persons and kin coexists with predatory relationships (Hall 2011, p. 100). As denoted previously, total abstention from harm and violence towards other persons is not practicable for heterotrophs, who must take life in order to live, and life itself is not possible without nutrient exchanges across multiple ecological scales. However, wanton harm, indiscriminate killing and exploitation are antithetical to 'right relations' and ecological integrity, and should be avoided as far as possible. It is also important to acknowledge the harm that we must do to some others so that we may live, which is to honour the kinship relations that enable our lives. The late Kenyan environmental activist and first Black woman to receive a Nobel Prize, Wangari Maathai, recounts the example of the Kikuyu tribe of south-central Kenya who have many rituals and practices for expressing gratitude for the lives that sustain them. Notable among these is the traditional practice wherein farmers always left a portion of what they harvested for local species to enjoy (2010, p. 19). Through prolonged and intimate cohabitation with their domestic companion species, Kikuyus grew to know their 'idiosyncrasies and responded well to their needs' (p. 23). Heartbreakingly, Maathai laments that these practices became increasingly rare after colonial and Christian contact, and following the spread of the 'cash economy', which turned everything into a commodified 'it'. As discussed above, when those whose lives we take so that we may live are regarded and treated as 'who's' with their own trajectories, proclivities and desires, rather than 'its', this profoundly alters the ethical responsibilities that we owe them. A 'who' cannot be for sale, and is always more than a mere means to an other's ends. The Honourable Harvest requires that the purpose for which we take life is 'worthy of the harvest' (Kimmerer 2013, p. 48). Thus, hunting a deer respectfully and with gratitude for the multiple people nourished and clothed by them is one thing; cutting the fins off of millions of sharks to use as a luxury ingredient in our food whilst the rest of their often still-living bodies are dumped into the sea, is quite another. Yet, Kimmerer reminds us that radical changes in orientation and ethical sensibilities are rendered possible through utopian imaginaries and pedagogies:

> Imagination is one of our most powerful tools. What we imagine, we can become. I like to imagine what it would be like if the honourable harvest were the law of the land today ... Imagine if a developer, eyeing open land for a shopping mall, had to ask the goldenrod, the meadowlarks, and the monarch butterflies for permission to take their homeland ... (Kimmerer 2013, p. 42)

Imagine, indeed. With such newfound ethical and political responsibilities, the developer in question would see the land as no empty or unoccupied piece of terrain but, crucially, a dynamic and living entity by way of the multiplicities of agencies who constitute it. In turn, the question of whether or not to raze part of a forest or build a hydroelectric power plant would require considerately engaging with relevant agencies who would be most affected by such decisions, and thus whose interests would need to be taken into account. It would involve carefully negotiating trade-offs, asking who bears the costs and who reaps the benefits in such ventures, not only for present but also future generations. However, respect and gratitude, though essential principles for ethically reorienting and guiding more ecotopian relations with others, are not sufficient on their own. Cultivating a wider culture of reciprocity is also important – asking not only what we can take under what conditions, but what we can give back and how we might otherwise enrich the lives of our terrestrial counterparts, both present and to-come (Kimmerer 2013, p. 53). Harvesting honourably requires that the taker as well as the giver be sustained and nourished, not steadily diminished. It calls for an orientation wherein terra and other terrestrials are *with* us, not *for* us (Le Guin 1985, p. 76).

History Is the Future and the Earth Itself

> How the land is treated – it is like using a stick and throwing the Earth up, taking all her valuables, and shaking it clean. (Pauline Whitesinger, Big Mountain Matriarch of the Hardrocks Chapter, Navajo Nation, cited in Benally & Denetdale 2011, p. 75)

As discussed previously, according to many indigenous ontological and ethico-political traditions and as seen in examples of indigenous-inspired ecotopian texts such as *Always Coming Home* and *The Fifth Sacred Thing* in Chapter 4, a territory is not merely a geographic space with solid and stable boundaries to be occupied, or measured in miles or kilometres. Territories are 'sentient landscapes' (Hall 2011, p. 106), dynamic assemblages constituted by multiple agencies, interdependencies and lifeways. One's territory extends 'as far as the list of interactions with those we depend on ...' (Latour 2021, p. 72). The late Lakota philosopher Vine Deloria Jr. argues in their seminal work *God Is Red: A Native View of Religion* (2003) that a key distinction between Indigenous and Western metaphysics revolves around the central importance of land to Indigenous ethics and cosmologies. In the latter's view, we are not merely *on* the land or country, like a billiard ball sliding across a pool table, nor a mere surface upon which we tread but something that we are *of*. Land or country, crucially, refers not to the territorial notion of a 'homeland' but rather encompasses 'humans as well as waters, seas and all that is tangible and non-tangible and which *become together* in a mutually caring and multidirectional manner to create and nurture a homeland' (Suchet-Pearson et al. 2013, p. 186). Land is a dynamic meshwork (Ingold 2006, p. 13) of human and other-than-human agencies linking past, present and future generations across time and in relation to a given geographic terrain. As such, land is alive, it is a who composed of other who's with responsibilities towards one another, and thus, like our terrestrial kin, it is (or ought to be) beyond private or individual ownership. It is precisely this intimate embeddedness with the land and its constituent ancestors that has served as a powerful force fuelling indigenous resistance against the omnicidal 'Now' (Ghosh 2021). In his essential book *Our History Is the Future: Standing Rock versus the Dakota Access Pipeline, and the Long*

Tradition of Indigenous Resistance (2019), Nick Estes chronicles in heartwrenching detail the centuries-long Native American resistance against settler-colonial appropriation of their lands, of which the Dakota Access Pipeline project serves as merely another recent example:

> #NoDAPL was also a struggle over the meaning of land. For the Oceti Sakowin, history is the land itself: the earth cradles the bones of the ancestors ... for others, however, the earth had to be tamed and dominated by a plow or drilled for profit. Because native people remain barriers to capitalist development, their bodies needed to be removed ... (p. 47)

History is not confined to a distant past, but is rendered contemporaneous through the land, the ancestors, and the enduring kin relations it harbours and is constituted by. Hence, in part, why so many indigenous struggles revolve around land rights and reassuming collective tenure over that which was stolen and ravaged by the 'violence of empire' (Yazzie 2018, p. 31). The industrial pollution and toxicity resulting from resource extraction continue to disproportionately perpetrate violence against the land, indigenous and other terrestrial bodies. Diné (Navajo) scholar Melanie Yazzie (2018) reiterates the central role that 'anti-capitalist decolonisation' plays in the diagnoses and utopian refusals of the deficient now amongst indigenous peoples like the Diné. Furthermore, Diné relational ethicoontological modalities centred on the vital responsibility of being a 'good relative' to all of one's relatives, other animals, plants and water included, serve as the essential foundations for their 'utopian' political mobilisations (Yazzie 2018, p. 34). Thus, yet another important dimension of their activism is to dismantle the artificial divisions between humans, other animals and the land imposed by settler colonialism, and to restore severed relations with ancestral and earth kin. In this vein, it's important to reiterate that in utopian struggles for justice and liberation, denunciation is not enough. As such, Yazzie alludes to the importance of annunciation alongside denunciation by stressing that resistance to the 'death drive' of extractive capitalism and settler colonialism in the form of the transgressive politics of anti-capitalist decolonisation must also 'function as a vehicle for *imagining* a *politics of life* [emphases added] that will ... secure a future for all our relations' (p. 31). Thus, the #NoDAPL occupation camps don't just serve as physical instantiations of 'Blockadia' (Klein 2015) against colonial-capitalism's ongoing assault on 'proper' or 'right' relations with terrestrial life; much like contemporary REAs discussed in the previous chapter, they serve as radical spaces of alterity wherein one can, however tentatively, (re)imagine and enact better worlds:

> At the camp, the smell of campfire brought us back to another world ... In the twilight hours, water protectors told stories and prophetic visions of a better world, not just in the past but one currently in the making ... (Estes 2019, p. 6)

In other words, these spaces serve the vital utopian function of reasserting the boundlessness of terrestrial potentiality, a reminder of worlds that once were and of new worlds that yet might be. Moreover, indigenous transgressive mobilisations against the deficient present and enactment of 'vibrant futurities' conducive to multispecies thriving (Yazzie 2018, p. 31) also come alive in their stories and in Indigenous speculative fiction. What might be the radical utopian potential of 'storytelling' and speculative fiction? To quote John Berger (2007) again:

> Story-time (the time within a story) is not linear. The living and the dead meet as listeners and judges within this time ... stories are one way of sharing the belief that justice is imminent. And for such a belief, children, women and men will fight at a given moment with astounding ferocity. This is why tyrants fear storytelling. (p. 96)

Berger was writing largely in reference to the ongoing Palestinian resistance against the Israeli settler-colonial occupation of their territory. However, this can also be applied to indigenous storytelling and speculative fiction, as we'll see below. As a result of the destructive legacies of European colonial expansion across – and appropriation of – the Americas, denotes Conrad Scott (2016), 'Indigenous literature ... tends to narrate a sense of ongoing crisis rather than an upcoming one' (p. 77, Saunders 2022). The climate and biodiversity crisis are not looming on some steadily approaching horizon – they're already here, indeed arrived some time ago, especially for the most vulnerable populations. However, far from acquiescing to a passive victimisation, many native peoples fervently engage in 'survivance' (Vizenor 2008; Dillon 2012), a 'narrative resistance to absence, literary tragedy, nihility, and ... an active sense of presence over historical absence ... [and] a continuance of stories' (Vizenor 2008, p. 1). This leads to an important question: how do indigenous onto-epistemological and ethico-political modalities, and the violent legacies of colonialism, capitalism and ecological ruination, influence their utopian impulse, and the content and structure of their visions of the better? Traditional or transcendental utopias in particular, as exhibited by much of H. G. Wells's speculative fiction, have been steeped in modernist notions progress and 'linear ideologies of advancement' (Darian-Smith 2016). By contrast, traditional Western ideals of linear progression from a deficient present towards a luminous futurity are decidedly complicated by many indigenous peoples' non-linear conceptualisations of time. Aboriginal peoples, for instance, experience all times simultaneously, so that one can more accurately speak of an 'everywhen' (Saunders 2022, p. 8). This matter is rendered doubly complex by their ongoing reckoning with colonial legacies of displacement and destruction of their cultures and traditional lifeways. Moreover, it's important to remember that the Western linear model of time itself is far from universal, but rather was imposed on many native cultures through colonisation:

> Settler narratives use a linear conception of time to distance themselves from the horrific crimes committed against indigenous peoples and the land ... but Indigenous notions of time consider the present to be structured entirely by our past and by our ancestors. There is no separation between past and present, meaning that an alternative future is also determined by our understanding of our past. (Estes 2019, p. 14)

The Anishinaabe have a concept of 'intergenerational time', which denotes a 'spiralling temporality' wherein the self is seen as walking through life alongside future and past relatives simultaneously (Whyte 2018, pp. 228–229). This sense of intergenerational temporality similarly suffuses Starhawk's *The Fifth Sacred Thing* and Le Guin's *Always Coming Home*, which delineate experiences of spiralling time characterised by irregularity, sometimes marked by cyclicality and at other times un-cyclicality (Whyte 2018, p. 229). Like the doings of our co-terrestrials and terra more generally, it is dynamic and at times wholly unpredictable. As such, in contrast to the prominent Western concepts of time denoted above, time is instead perceived and experienced as disjointed and variegated. Indeed, humans operate along decidedly different timescales than, for instance, insects, some of whom (i.e. mayflies) only live for a day. Greenland ice sharks, on the other hand, can live longer than 250 years, whilst trees appear to be the longest-lived organisms on earth. The average oak tree can live up to an astonishing 1,000 years, whilst a 'Great Basin bristlecone pine' in the Western United States currently estimated to be over 5,000 years old might be the oldest living terrestrial to ever inhabit terra (Chesterton 2017). Even a mountain, which can almost be considered immortal due to its seeming immobility and enduring presence in relation to ourselves, in the words of John Berger, 'never repeats itself. It has its own timescale' (2007, p. 50) – at least to those who are familiar with it. However, importantly the notion of spiralling time does not 'foreclose linear, future thinking' ... rather, it is 'a dialogical unfolding that also has, in a sense, forward motion that can be both predictable and irregular' (Whyte 2018, p. 229). Therefore, in such alternate temporalities, pasts, presents and futures 'flow together like currents in a navigable stream' (Dillon 2016, p. 345). The Australian Aboriginal concept of 'Dreaming' denotes a similarly timeless, uninterrupted flow of action by individuals in collaboration with communities across eternity, marking a further deviation from Western linear temporalities (Dutton 2010, p. 247). The above yields a perception of time – like 'reality' as such – that is decidedly fluid, devoid of hyper-separations, and open-ended.

The aforementioned highlights the need to continue the essential work of decolonising the utopian canon. Often and understandably, the seminal works of Plato, Thomas More, William Morris, H. G. Wells and others come to mind when we think of the field. Hence Kumar's parochial assertion that there is no discernible utopian tradition beyond the West (1987, p. 19) because other cultures purportedly lack the foundational myth of the Golden Age – i.e. the Garden of Eden, Atlantis and Arcadia – that has been so formative to the development of the Western utopian tradition (Dutton 2010, p. 230). However, the utopian mode and impulse is far from an exclusively Western phenomenon (Bagchi 2016, 2020). Barnita Bagchi (2016), for instance, documents the (re)surgence of utopian fiction, activism and strivings for social transformation under British colonial rule in India and in other South Asian contexts. From feminist utopian writers and social reformers such as Kashibai Kanitkar arguing for women's education and emancipation from patriarchal domination to Raindranath Tagore, a Bengalese Nobel-Prize-winning writer and utopian social reformer who created a cosmopolitan utopian community in Santiniketan in rural Bengal (present-day India). Tagore's grounded utopian space centred around education as a key vehicle for social transformation, where people could engage in dialogue in 'open classrooms under trees' beyond the confines of 'narrow nationalism' and the colonial formal classroom (Bagchi 2016, pp. 210–212). In 2008, a 'concrete ecotopia' in the form of the Zero Budget Spiritual or Natural Farming movement (ZBSNF) of farmers sprouted in Wayanad, Kerala seeking to effect a 'paradigm shift in agrarian human-environmental relations' (Münster 2015, p. 241). Crucially, the ZBSNF farming system aims to promote farmers' autonomy from exploitative market forces[3] and practice chemical-free agro-ecological methods centred on an understanding of soil as a dynamic and living component of 'nature' (Münster 2015, p. 243). Native desi cows play an especially important role in enhancing soil health and fertility through their assistance in the production of fermented jivamrta. The latter is prepared by mixing the cows' urine and dung (which allegedly have a much higher concentration of micronutrients than the urine and dung of hybrid cows) with pulse powder and a sweet component such as fruit with water, and then left to ferment for several days (Münster 2015, pp. 243–244). When applied to the soil, jivamrta does not act as a fertiliser but rather 'enhances microbial activity in the soil by attracting earthworms, which in turn bring nutrients from lower levels of the earth to the topsoil' (pp. 243–244). Münster (2015) notes that the farmers who have utilised jivamrta have unanimously reported an array of financial, productive, health and ecological benefits, thereby obtaining 'a sense of an agrarian future in a situation of crisis and structural futurelessness' (p. 13).

Western society is just one of 'multiple modernities' (Bagchi 2016), albeit a particularly pervasive one. Similarly to the concept of the 'pluriverse' (De la Cadena & Blaser 2018; Escobar 2020), this notion is crucial for challenging monolithic conceptions of modernity, temporality and in this case the utopian imaginary. Though Western-capitalist modernity has indeed

---

3   Through, among other things, a shift from monocropping cash crops to mixed cultivation.

operated as a hegemonic force in many ways, this is a world of many worlds, spatial-temporalities and dynamic understandings of 'the good life'. Many of the South Asian utopias examined by Bagchi (2016, p. 217), for instance, evince features characteristic of the 'critical utopias' discussed in Chapter 3 in their emphases on contingency, dynamism and an aversion to stasis. We might come to better understand such 'intercultural imaginaries of ideal societies' (Dutton 2010, p. 225) by more concertedly engaging with groups traditionally consigned to the margins of modernity, whose unique histories, origin stories, ways of knowing and being in the world often yield rather different ways of imagining 'the better'. However, in doing so, it's important to remain wary of the tendency to project Western definitions of utopia onto 'non-Western' representations of the desire for alternative worlds, thereby failing to 'challenge the epistemological foundations of utopia as a Western way of imagining the ideal society' (Dutton 2010, p. 227). Times of upheaval and crisis provide fertile ground for the resurgence utopian imaginings and strivings, and this was no less the case in the wake of the horrors of colonialism, out of which emerged the Native American Ghost Dance movement in the late 1880s. This upsurge of utopian desire in defiance of the dark lived realities of the continent's native populations was inspired by the prophecy of a Paiute holy man known as Wovoka, who envisioned a time beyond the present marred by colonial violence and dispossession (Estes 2019). In this better world, human relatives, Buffalo nations and other relations who'd been exterminated by colonial invaders would once again roam the earth in a renewed abundance (Estes 2019, p. 122). Wovoka's prophecy foretold that, at some time in the near future, a 'cataclysmic event – such as an earthquake or whirlwind – would wipe the United States off the surface of the earth' (Estes 2019, p. 122). The Ghost Dance, along with Le Guin's *Always Coming Home* and Starhawk's *The Fifth Sacred Thing* which were suffused with purposive more-than-human agencies, further highlight the prominent role that other-than-human beings, spirits and other marginalised entities play in indigenous cosmologies, utopian imaginings and fiction, often appearing in the latter as key protagonists rather than mute backdrop (Dillon 2012). Such worlds of intergenerational temporalities and expansive kinship networks feature prominently in the anthologies discussed below.

Love After the End

Joshua Whitehead's moving anthology *Love After the End* (2020) features speculative stories by two-spirit and indigiqueer authors seeking to imagine and explore the possibilities of life and love in and beyond the 'end times' of the eco-dystopian present – on terra and beyond on other 'relative' planets. Liminal beings of all kinds are featured as central characters – AI rats and motherships, synthetic humans, and ihkwewak, the Plains Cree word for 'two-spirit' people thought to serve the crucial role of intermediary between spirits and the physical world (Calderon in Whitehead 2020, p. 103). The stories are situated in a range of apocalyptic and postapocalyptic scenarios within a wider backdrop of a deeply damaged terra in its final death throes due to runaway climate chaos or, in the case of Mari Kurisato's 'Seed Children', scorched by a dying sun (Whitehead 2020, p. 138). In response to such cataclysms, many – especially the super wealthy and powerful – have either bought floating doomsday cities in the middle of the ocean away from the turmoil on land (p. 119), or fled to colonies on Mars, the moon and beyond. Many of those 'too poor or too unskilled to secure passage off of the Withering earth' (Kurisato in Whitehead 2020, p. 138) were left to die or carve out a meagre existence on an increasingly inhospitable planet. As the title suggests, the stories variedly explore and grapple with the timely question of, what comes after 'the end' – of a given system, of life on terra as we knew it, of relations with cherished others? How to create liveable worlds amidst blasted landscapes (Tsing 2012) marred and scarred by power inequities, exploitation and widespread degradation? How to build caring relations with strange new creatures and planets?

Kinship is central to indigenous ethics and politics in the 'Now', as well as in the utopian imaginaries in *Love After the End* (2020). As discussed, kinship involves not only other humans but, crucially, other beings as well – from plants and animals to mountains and rivers, and Nimama Aki – Mother Earth. Grief and mourning over the death of Nimama Aki, and a profound reluctance to escape to post-terrestrial worlds, is rife in stories such as Jaye Simpson's 'The Arch of the Turtle's Back' (p. 69) and Adam

Garnet Jones's 'The History of the new world'. In the latter, protagonist Emma refuses to leave a dying Earth with their family to find refuge in the New World, a twin planet with seemingly endless natural riches, no history 'except that which the people brought with them' (p. 44), and thus ripe for the taking. Emma's reluctance stems partly from a refusal to abandon Earth kin and a determination to (re)build habitable spaces on terra. Emma also wishes to break cycles of colonial violence that have ravaged the earth and their ancestors by not participating in their export to the New World. As it turns out, and as might have been expected, the New World is far from 'empty' but is populated by other species who are 'some kind of people' with language and a history (p. 45). Eventually Emma's partner decides to join other colonists in the New World and Emma stays behind with their daughter Aseciwan, making their way through empty streets reclaimed by animals and plants, 'a twilight time, after the unchecked human explosion, and before whatever came next' (p. 58). Eventually the New World falls apart, and one catches a glimmer of hope in the 'High Law' newly formed on Earth, which denotes 'shared responsibilities between people and all our relations: the ones that walk on four legs, the ones that swim, the ones that soar in the air, the ones with leaves and branches, our grandfathers the mountains and our grandmothers the waters' (p. 60). Of course, this doesn't mean that 'cycles of war and peace, love and heartbreak, hunger and feasting' cease, but that balance in the sense of the 'Honourable Harvest' is an ongoing, ceaseless undertaking (p. 60).

In 'The Arch of the Turtle's Bach', after much painful deliberation, protagonist Ni and some of their kin eventually agree to part with a soon-to-be uninhabitable Earth and head towards a new colony on the moon for the sake of future generations. However, in casting their gaze to the new horizon of possibility, Ni asks the attendant to their cryogenetic pod how one is to build a relationship with their new home planet, to which the attendant responds, 'I would assume like all consensual relationships: we ask them out' (p. 76). A difficult relationship to establish whilst otherwise in the midst of colonising, appropriating and terraforming other worlds. Yet such ethical dilemmas remain at the forefront of many of the stories' deliberations about the ethics and politics of making new homes on and beyond terra – what and who do we leave behind? Who might be affected by our arrival and settlement, and how? What is lost in the process?

Love – along with respect, care and attentiveness – is an integral component of kinship relations, the essential foundation on which they rest. With love and careful attention, we might come to understand what even distant 'others' such as plants need, because, though they cannot speak, they 'show us what they need' (p. 125). Personhood is 'manifested in communication' (Hall 2011, p. 108), and (human) speech is merely one amongst a vast range of communicative modalities. As alluded to by its title, throughout *Love After the End,* love is a central, cross-cutting theme – a concept or force which helps human and other-than-human people (re)build in and after the apocalypse. In Nathan Adler's 'Abacus', an AI rat named Abacus and human boy named Dayan fall in love and escape from a space station to a colony of runaway AIs on Io (p. 34), one of Jupiter's moons and the most volcanically active body in our solar system. Here kinship isn't limited only to terra or Nimama Aki:

> Dayan: 'If earth is our mother and the moon is our grandmother – what does that make Io? What does that make Jupiter?'
>
> Eva [Dayan's mother]: 'Relatives too. Aunties. Uncles. Cousins.' (p. 23)

As philosopher Michael Hardt observes in his joint work with the political philosopher Antonio Negri *Multitude: War and Democracy in the Age of Empire* (Hardt & Negri 2005), the modern concept of love has been largely restricted to the private domain of the nuclear family. For Hardt, restricted love or 'Love of those like yourself has destroyed the possibility of love as a more generous and positive political concept' (Schwartz 2009, p. 812). Love of the same is what, in part, fuels nationalistic and far-right movements. Rather, Hardt calls for a shift towards the latter, far more 'unrestrained' understanding of love as a political concept that does not stop at the familiar or self-same. Crucially, this love, in opening itself towards a love of the 'different', of the 'stranger', harbours decidedly transformative power by providing a basis for the collective construction of a new society (Schwartz 2009, p. 812–816). This is the modality of love that *Love After the End* variedly gestures towards and explores throughout – love as a powerful act of defiance and resistance against unjust and oppressive conditions, and as that which opens the door towards new possibilities: 'They kissed like the world was ending, but really, wasn't it already over, and perhaps within this kiss lay the new beginning?' (Calderon in Whitehead 2020, p. 109). The utopian potentiality of love in its positive political modality is similarly explored briefly in Yanis Varoufakis' recent utopian work *Another Now: Dispatches from an Alternative Present* (2020). Herein, a central character Iris, who's charged with casting a self-critical eye on the utopian alternative to the present one marred by the ravages of neoliberal capitalism, remarks:

> Falling in love is one of the greatest acts of resistance against Thatcherism's oeuvre!' she would declare. 'In an era when being in control – of one's stocks, inventories, workers, timetable – is valued above all else, falling in love means surrendering control to an 'other'. It threatens the foundational ideology – of exchange value, of individual agency and self-determination – of financialised capitalism. (p. 162)

Like individuals and species in all of their uniqueness, whatever it is that two lovers are 'giving on another can never be itemised or quantified', and similarly never recouped; it is wholly singular (p. 166). Recalling Chapter 5 on the love and deep kinship bonds exhibited by REAs in relation to Earth kin, unrestrained love – for innumerable, singular Earth others – in its resistance to full approximation, understanding or control serves as a powerful act of defiance against oppressive hegemonic systems. It resists the treatment of others as mere resources or quantifiable and exchangeable commodities. Polyamory and romantic love for someone of the same sex constitute similar acts of resistance against the violent and exclusionary dictates of the heteronormative regime that pathologises deviations from the heterosexual norm, another major theme running throughout *Love After the End*. Of course, love and kinship are not without their difficulties and blind spots. For instance, kinship 'is just as much about hating each other and messing each other up as it is about loving each other' (Pyle in Whitehead 2020, p. 83). In other words, love does not entail the absence of conflict, nor is it static and unchanging. Moreover, Inawemaagan (the Ojibwe word for kin) is often a 'two-sided coin' … 'you gotta ask yourself, who is being excluded here?' (Pyle in Whitehead 2020, p. 87). This is an important question that we'll explore below and in the final chapter. Other than love, care and kinship, how else does one survive an apocalypse? As the final section in Kai Minosh Pyle's story 'How to Survive the Apocalypse for Native Girls' suggests, 'the only way to survive the apocalypse is to make your own world' (p. 94) in defiant resistance to the dystopian surroundings without. Recalling Starhawk's *The Fifth Sacred* Thing, you build liveable spaces amidst the ruins that nurture love and care, thereby keeping hope and the possibility of better worlds and relations alive (1993, p. 238).

This All Come Back Now

> In the quiet calm, in conversation with Country, I hear the whispers of another way of being, and that is the call that I must follow ...
>
> – Laniyuk in Saunders (2022, p. 219)

Mykaela Saunders, a queer, working-class member of the Tweed Goori community, describes her edited collection *This All Come Back Now* as the first anthology of speculative fiction of, by and for Aboriginals and Torres Strait Islanders. Its title, taken from its first entry 'Muyum, a Transgression' by Evelyn Araluen, refers to the tendency of things to come back – i.e. waves to their rivers, and losses that we attempt to recover or revive, or which come back to haunt us (p. 14). Two stories in particular stood out to me by way of their foregrounding of some of the key themes discussed throughout – i.e. of recalcitrant other-than-human agencies, and the challenges and urgency of kin-making during and after times of crisis. Ellen van Neerven's provocative story 'Water', set in a near-future Russell Island, recalls Jones's 'The History of the New World' in Whitehead's *Love After the End*. The Australian government are in the process of creating an 'Australia2' for relocating Aboriginal people, and must first displace the territory's original inhabitants. The story's human protagonist Kaden, a 'cultural liaison officer', is thus sent to prepare the 'sandplants', the territory's human-plant hybrid inhabitants, for relocation (p. 223). At first, Kaden is struck by 'how startlingly human-like they are, and how alarmingly unhuman they are' (p. 226); though, after becoming better acquainted with the sandplants through her budding relationship with their leader Larapinta, she engages in a series of critical reflections and mediations on human/nonhuman boundaries, interspecies communication and affinities, as well as differences, particularly across two different taxa often presumed to be worlds apart. Kaden enquires if Larapinta is a someone or a something (p. 244), whether she can think or if 'it's just processes' (p. 237), and more generally what constitutes the 'human' and other-than-human (p. 238). After discussing mining and islandising activities for Australia2, Kaden remarks, 'I thought you weren't political', to which Larapinta replies: 'There isn't anything that isn't political' (p. 227), politics traditionally being considered an exclusively human activity. The story closes with Kaden and Larapinta exploring a nascent romantic connection, in response to which Kaden muses: 'How much of what it means to be human will sway deep in my mind like a ship?' (p. 244)

Saunders' own short story 'Terranora', set in her home Tweed community, examines the difficulties and complexities of sustaining caring relations in a climate-ravaged world. Care isn't easy; it can often be emotionally taxing, as well as time and resource-intensive during the best of times. Moreover, care is a double-edged sword – care for one's own kin and community always at least implies the exclusion of outsiders. The story centres around a tight-knit Goori clan in a territory called Terranora who take care to follow community laws set forth by elders so that there are enough resources for its members/kin in a country 'being killed' by 'centuries of pollution and overfishing' (p. 253). The story arch is interspersed with descriptions of agentic other-than-human inhabitants of Terranora, announcing their presence and intentions, such as 'every resident of the estuary' beginning to 'sing or scratch or bark or buzz according to their on cultural ways' at sunset (p. 259). Though Nan Jack, one of the community elders, describes the Goori clan as temporary guests in a 'diverse community of life – a complex network of relationships, and an elegant web of reciprocity' (p. 254), she also highlights the need to protect this fragile network from others who might seek to take advantage. This is put to the test with the arrival of an older woman seemingly suffering from Alzheimer's, who is first befriended by a character named Skinny Kel and later by the protagonist Valmae. Initially the elders are reluctant to welcome the old woman, as she would need additional care which would 'take time and food away from my community' (p. 263). However, Valmae challenges the limits of the Goori Clan's hospitality and community boundaries by volunteering to adopt the older woman, emphasising the ethical importance of making space for newcomers. With mounting climate perturbations and resulting mass mobilisations in the decades to follow, we would do well to critically interrogate previously held assumptions about who belongs where, and who might be welcomed as a potential new kin member.

Indigenous Futurisms

> We are the descendants of Little Thunder, who witnessed the massacre that cleared out the Great Plains to make way for the cowboys, cattle, and industrial farms. We have seen the great trees felled, the wolves taken for bounty, and the fish stacked rotting like cordwood. Those memories compel us, and the return of the descendants of those predators provoke us to stand again, stronger, and hopefully with more allies. We are the ones who stand up to the land eaters, the tree eaters, the destroyers and culture eaters.
>
> – Laduke (2015, p. 3)

Yakthung Subash Thebe Limbu from Eastern Nepal describes himself as an 'indigenous artist imagining futures', deploying his creative practice to explore and give impetus to an 'Adivasi futurism'. Inspired by Indigenous and Afrofuturisms seeking to redefine themselves, technology and the future from liberated standpoints, Subash envisions Adivasi futurism (Adivasi is a Nepalese word for 'indigenous', used more specifically in reference to Indigenous peoples across the Indian sub-continent) as a space wherein 'Adivasi artists, writers, musicians and filmmakers can imagine and speculate future scenarios from their own perspective, where they have agency, technology, sovereignty and also their indigenous knowledge, culture, ethics and storytelling still intact' (Limbu 2020). Indigenous mobilisations for better worlds in the form of Standing Rock and Idle No More for the protection of indigenous rights, water and sky persist in defiance of multiple 'ends' imposed by colonial-capitalist annihilation. As such, indigenous speculative fiction in particular tends to be grounded within a wider backdrop of extant and ongoing crisis, not a crisis (climate, biodiversity) that is merely imminent. Their ecotopian sensibilities are intimately shaped by their onto-epistemological and ethico-political modalities which are often considerably distinct from their Western counterparts. Crucially, many cultures exhibit non-hierarchical ontologies which underlie their thoroughly collectivist and egalitarian modes of relationality with other species and terra. The latter, importantly, are regarded as kin, not as external or inferior entities to be controlled or exploited at will. A radical awareness of and respect for the interdependent kin relations that sustain us helps discourage their overexploitation and the systematic annihilation of other terrestrials in the business of living (Hall 2011). In indigenous metaphysics, 'everywhen' time and the land are dynamic, intergenerational assemblages of kinship relations, wherein land is understood as that which you are *of* rather than *on*. The above, in combination with the general absence of an ontological divide between a 'mundane' terrestrial realm and a transcendent order wherein salvation is to be found, as well as linear or teleological conceptualisations of historical time as an inevitable march towards 'progress', engender visions and strivings towards the better that are firmly rooted in terra, rather than in a distant planet or afterlife.

In other words, as with the contemporary REAs discussed in Chapter 5, what emerges is a similarly terrestrial utopianism wherein the better is to be brought about by (re)establishing respectful relations with the land and terrestrial kin in the here and now (Deloria 2003). Of course, this does not mean that indigenous utopian strivings don't sometimes gesture towards distant futurities or other planets. As Subash further reflects, one key aim of Indigenous and Adivasi futurisms is 'to re-imagine ourselves as not only the storytellers of the past but also as creators of interplanetary and interstellar civilisations of the future' (Limbu 2020). However, as seen in Whitehead's *Love After the* End and Saunders' *This All Come Back Now*, orientations towards the beyond take the form of relations between relatives, rather than as disembodied entities to be colonised and terraformed to suit human ends. In other words, if we are to cast our gaze towards the vast expanse of potentiality that is the beyond terra in search of more liveable futures, great care must be taken not to extrapolate or reproduce those exploitative relations and imaginaries that have imposed so many violent ends on terra. Indigenous peoples' place-based ontological and ethical orientations, in the words of Dene scholar Glen Coulthard, continue to serve as 'the radical imaginary guiding our visions of a just political and economic relationship with non-Indigenous people and communities based on principles of reciprocity and mutual obligation' (2010, p. 81). Similarly, for sociologist Max Ajl (2021) indigenous claims to land harbour considerable utopian potential in their implication of a reversal of primitive accumulation (p. 379). More so, the ethical practices and traditions of many indigenous groups, which typically emphasise respect, care, gratitude, reciprocity, attentiveness and love, harbour much ecotopian potential for a radical (re)design of our relations with our terrestrial kin. This isn't to say that care and intraspecies cohabitation will always be easy or devoid of conflict, particularly amidst increasingly precarious terrains, but they offer helpful foundations for learning to be better relatives.

CHAPTER 7

For the Love of Kin: Terrestrial Ecotopias in and After the End

> But every respite, every little intermission in business-as-usual is a reminder that a world – not another world, this world – might still be possible.
>
> – Malm (2021, p. 130)

Another world here on terra remains an enduring, inexhaustible possibility. Hence, hope as utopian life-affirming desire and praxis, springs eternal, particularly during dark times, spurred by an ardent rage against concrete intolerable injustices and a passionate yearning for liberated worlds. It functions as a vital antidote to the 'ethicide' of oppressive and destructive systems like capitalism, anthropocentrism, Eurocentrism, patriarchy and the heteronormative regime which 'kill ethics and therefore any notion of history and justice' and, worse, 'bury imaginative spaces' (Berger 2007, p. 83). And these are dark times, indeed – present and coming generations of terrestrials will navigate an increasingly damaged world ravaged by fires, floods, storms super-charged by warmer seas and temperatures, chronic resource shortages, ever greater portions of the globe rendered virtually uninhabitable, and resulting socio-political turmoil. Climate apartheid and climate gentrification (Taylor et al. 2022) are likely to proliferate as wealthy private and state actors capitulate on mounting crises to enhance their own relative power and security. In the wake of Russia's invasion of Ukraine and the ensuing global energy crisis, countries around the world are doubling down on new oil and gas infrastructure (Climate Action Tracker 2022a), with the UK, under the ill-advised direction of the former Business and Energy Secretary Jacob Rees-Mogg, recently lifting its former ban on fracking (Walker & Algretti 2022). Record heat shattered global records in the summer of 2023. In the face of such harrowing prospects, and the seeming foreclosure of a liveable futurity, a descent into nihilistic despair, though ill-advised, is not uncommon or unreasonable. Some like the writer Roy Scranton in his work *Learning to Die in the Anthropocene: Reflections on the End of a Civilisation* (2015) suggest that it's already too late to avert total collapse and that we ought to embrace the inevitable end. However, along with others (McKinnon 2014; Wallace-Wells 2019; Malm 2021) this book has been in many ways an attempt to proclaim the dangers and ethical harms effected by such defeatist resignations. For one, they begin with the erroneous assumption that the perceived improbability of a world conducive to interspecies flourishing translates into its impossibility; worse than effectively denying the pluralistic and always-unfolding nature of 'reality' whichever defies attempts at absolute approximation, such a politically fatalistic stance extinguishes desires to resist the profound injustices of the present. As Andreas Malm (2021) observes:

> If someone seeks to affect the ways of the world by acting in one way rather than another, it must be because she holds an outcome to be desirable and wants to contribute to its realisation. If she merely wished to confirm the most probable *outcome on account of its high probability*, she would have no reason to act at all. (pp. 141–142)

The very foundation of hope is radical uncertainty, which renders hope and utopian strivings possible (McKinnon 2014; Solnit 2016). Conversely, where there is certainty about a given outcome, there can be no hope. At this present juncture, we stand at a crossroads with a range of potential pathways before us – ranging from 2 to 3 degrees and potentially higher by 2100. The IPCC's Sixth Assessment Report (2022) is unequivocal that some 'irreversible' impacts on the adaptability of human and natural systems from climate disturbances (p. 36) have already been locked in. The dream of keeping 1.5 alive indeed looks to be in its death throes (Matthews & Wynes 2022), with no country's climate crisis mitigation plans (for which there is reliable data) currently compatible with a 1.5-degree world.[1] However,

---

1 The Climate Action Tracker's traffic light ranking system sorts countries' climate mitigation plans (39 countries in addition to the EU, covering 85% of global emissions) from '1.5C Paris Agreement compatible' to 'critically insufficient'. As of

swift and sweeping action can still reduce the likelihood of nightmarish 3- and 4-degree worlds by 2100. In other words, *how much* warming occurs, *how much* suffering present and coming terrestrials will endure, how many species are consigned to the abyss of non-existence, and whether we initiate better rather than worse futures depends on what actions we take *today*. Of course, future scenarios do not follow on in a linear fashion from political will instantiated now but, rather, result from 'infinitely complex relations and conflicts and mediations'[2] (Berardi 2017, p. 14). Yet, by doing nothing, one affirms the probability of the worst possible scenarios; the fact that neither of these outcomes is guaranteed, and that the future is never written, is what provides fertile ground for action. Inaction is the opposite of hope; as Greta Thunberg wonderfully puts it, 'Hope doesn't come from inaction and empty promises that everything will be alright, they say trust us we are doing everything we can, that is not hope, hope is this, *hope is us*, the people, hope is when people gather to make change' (Fridays for Future protest in Milan, October 2021). Terra, this dynamic assemblage featuring many worlds simultaneously, is quite literally our only home. We share this singular habitable zone and a finite earthly mortality with countless other terrestrials, though we operate across vastly different timescales. Whatever lies beyond in the post-terrestrial realm is literally beyond our direct experience (Latour 2021). Thus, it is in the here below where we ought to direct our efforts for a liveable Not-Yet, a responsibility we owe our terrestrial relatives. This is what the utopian imaginary in its myriad manifestations is all about – a fervent refusal to acquiesce to injustice, oppression and exploitation, or to accept 'what is' as necessarily so. In this closing chapter, I discuss the ethical significance of loss and life amid times of extinction, which serve as a further impetus for ecotopian

---

September 2022 most countries' pledges fall within the 'Highly insufficient' and 'Insufficient' categories (Climate Action Tracker 2023).

2   Although Berardi (2017) seems to designate human consciousness as the primary source of the 'potency' that actualises the myriad possibilities inscribed in the present (p. 17). Whereas, along with others in the critical-posthuman tradition I maintain that humans along with other terrestrials possess varying degrees of power and 'potency' in different situations. Humans are far from the only ones who act, plan, conspire, etc. in order to instantiate a range of possible futures.

politics, and the intricacies, challenges, wild potential and urgency of a terrestrial ecotopianism for helping us build beyond 'the end'.

'A pocket universe gone ...'

> It was so hard to imagine that a mind could be gone. All those thoughts that you never tell anyone, all those dreams, all that entire pocket universe gone. A character unlike an other character, a consciousness.
>
> – Robinson (2020, p. 499)

A pressing imperative to enact more liveable worlds in the here and now stems from the ethical responsibilities that we owe our invaluable and quickly vanishing terrestrial kin. The ongoing sixth mass extinction as an often quiet 'systemic process of loss' (Wolfe 2017, p. 1), spurred by growth-oriented capitalism and Anthropocentric worldviews which instrumentalise and de-animate earth kin, continue to end worlds daily. What does the loss of a species as an evolutionary lineage in the wider fabric of terrestrial life signify, particularly during times pervaded by the mounting absence of other-than-human life forms? Why do these losses matter, when we know that the previous five mass extinction events have been followed by 'swift resurrections' of novel life forms (Rolston & Regan 1988, p. 187)? What do we lose when we lose a forest? Ursula Le Guin's heart-wrenching novel *The Word for World is Forest* (1972), which explores the corrosive violence wrought by colonial conquest on the vanquished and victors alike, offers a helpful entry point. For the Athsheans, inhabitants of the richly forested world Terra that has found itself in the crosshairs of Earthly colonial ambitions, 'the word for *world* is also the word for forest' (p. 60). For those who live in, constitute and depend on Terra's forests, they don't exist outside of or beyond a separate 'human' realm; they are their whole world – their source of solace, nourishment, and shelter, that which envelops them (Maathai 2010, p. 53). If the forests go, their fauna and floral symbionts go as well. Mass deforestation driven by neocolonial and extractive capitalism doesn't just result in the loss of individual trees but of entire worlds in the form of singular floral, faunal, and mycorrhizal assemblages containing diverse timescales, cultural traditions, hunting practices, symbiotic partnerships, and the like. This is why reforestation – the practice of planting trees to combat climate change and boost ecosystem resilience – though laudable, should only ever be pursued as a last resort. The razing of a forest effects the irrecuperable loss of its singular assemblage, which no amount of replanting can ever bring back or replace. Far preferable is what environmental scientist William Moomaw and others (Moomaw et al. 2019) refer to as proforestation, which entails leaving mature forest communities intact as much as possible. A further benefit to this practice is that mature trees are better at sequestering carbon than young ones.

Environmental philosopher and 'extinction studies' scholar Thom Van Dooren enquires: 'What does it mean that, in this time of incredible loss, there is so little public (and perhaps also private) mourning for extinctions?' (2014, p. 140). He suggests by way of a response that at the core of this seemingly pervasive apathy, which also featured as a core diagnostic theme in literary and social movement ecotopias, is an inability to grasp the multiple connections between ourselves and Earth others, an orientation that can partially be explained by the still dominant paradigm of human exceptionalism (Plumwood 2009; van Dooren 2014, p. 141). REAs and many indigenous cultures, in light of their deep kinship bonds with other species and the wider earth system, categorically refute and strive to dismantle hierarchical and dualistic constructions of the human/other-than-human relationship. Lacking such human-exceptionalist views and bolstered by high valuation of – and a sense of intimate entanglement with – the wider terrestrial world, REAs feel compelled to go to extraordinary lengths to prevent their further decline.

> For each time, and each time singularly, each time irreplaceably, each time infinitely, death is nothing less than an end of the world ... Death marks each time, each time in defiance of arithmetic,[3] the absolute end of the one and only world ... the

---

3 As with death, life too defies arithmetic in Le Guin's *Always Coming Home* (1985): 'But numbers are wrong. They are in error. You don't count scrub oaks. When you can count them, something has gone wrong' (p. 240).

end of the totality of what is or can be presented as the origin of the world for any unique living being, be it human or not. (Derrida 2005a, p. 140)

In Chapter 5, we saw that the loss of singular earth kin as experienced by many REAs result in no less than the severing of a social bond; this entails not merely the end of *a* world but *an* end of *the* world in the sense denoted above (Derrida 2005a). Each living being – from mycelia to California redwoods and sperm whales – represents a singular origin of existence, a unique patterning that constructs and interrupts our one and only world (our shared terrestrial existence). Though we share in common with earth others a finite mortality and embodied earthly habitation, these singular worlds can never be wholly appropriated by us and, crucially, can never be recouped once lost. This is what distinguishes a tractor from a sperm whale; The latter alone (or, at least more significantly) represents a singular, irreducible and therefore irreplaceable patterning in the world in the form of learned experiences, aptitudes, memories (Fritjof 1996). This is why such entities exert an ethical pull on us, and why we owe them certain responsibilities. When an entire species passes from the world (van Dooren 2014, p. 4), what transpires is not merely the epistemological or quantifiable loss of biodiversity, nor the mere departure of a 'fixed' population of organisms extinguished by the death of its last living member, as with the death of Sudan in March 2018, the last living male northern white rhinoceros, leaving his daughter and granddaughter as the final remnants of a doomed lineage (van Dooren 2014, p. 39).[4] Species are 'embodied intergenerational achievements' (van Dooren 2014, p. 27), whereby individuals exist as singular entities situated in complex co-evolutionary spatial-temporalities that extend from their past descendants on through the now and towards futures of infinite potentiality and diversity (in a Blochean sense). In other

---

4   Assisted reproductive techniques – i.e. the use of 'induced pluripotent stem cells' (iPSCs) from genetically diverse individuals of the functionally extinct northern white rhinoceros and from closely related southern white rhinoceros- are being explored in order to help repopulate the species (Korody et al. 2021). Nevertheless, as discussed in Chapter 6, such techniques don't restore the unique individuals and their relations with others. Who or what is brought back through such interventions?

words, a single individual spans inconceivably vast spatial-temporal distances as the 'knot' in time uniting all those who have preceded it, their uniquely embodied adaptations in the 'Now', and those who might branch forth thereafter. Or, as environmental philosopher Holmes Rolston puts it, unrelated causal and indeterminate lines converge to produce each individual as a 'one-time event' wherein no two coyotes inhabiting the same region are alike (Rolston & Regan 1988, p. 183).

What's more, such loss effects a permanent disruption of their intricate entanglements with myriad other unique beings (van Dooren 2014) situated in particular evolutionary communities, taking place within – and exerting immeasurable impacts upon – relations with others; it is the loss, in effect, of ethical-political relations, which is what makes extinction a fundamentally collective event. As such, to mourn and resist further loss is a profoundly political act (Stanescu 2012). It is the sheer enormity of this kind of loss, exacerbated by our common though differential complicity in it, that exerts a powerful ethical pull on us (van Dooren 2014, p. 39). Kinship ties to – and grief over – Earth others do not vanish along with their departure from terrestrial existence. From an evolutionary perspective the very capacity to grieve the loss of an *other* is a biosocial achievement developed through millennia of co-evolution and often intimate cohabitation with others (Archer 2003; van Dooren 2014). Similarly, from a psychological perspective, grief is no mere fleeting emotion but a complex and protracted process by which one engages and comes to terms with loss (Lazarus 1999, p. 656). Traditional psychoanalytic accounts of mourning tend to advise that we relive and then relinquish our memories of the dead (Freud 1984). However, for Derrida (1995a) the death and mourning of *an other* (particularly beloved kin), and extinction more broadly, are distinctly ethico-political phenomena because they are inextricably constitutive of self-other relations; that is, both life and death are thoroughly relational affairs that implicate multispecies worlds or assemblages (Dastur 1996; van Dooren 2014). Mourning in the Derridean sense entails not an abandonment of the departed, cherished other but an active affirmation of their unsubstitutable 'otherness', of our enduring connection with them, and our broader connection to some sense of a beyond (Derrida 1986, p. 85; 1995; Kirkby 2006, p. 464). Herein there is no possibility of permanently severing ties to the dead in order to reconnect to the world of the living (Freud 1984) because death is the very 'concrete structure of the living present' (Derrida 2016, pp. 70–71). In other words, the dead are simultaneously within and constitutive of us as well as beyond us (Dastur 1996), as our own speech and life-worlds are always laced with traces[5] of those who have lived before us (Kirkby 2006, p. 467). As such the border between life and death always remains 'open and ultimately interminable' (Derrida 1995a, p. 78), not unlike for many indigenous cultures (see Chapter 6). Similarly, for Bloch (1986) the tendencies of materiality are forever precariously suspended between better and worse modes of being, between the deficient 'Now' and endless potential 'Not-Yets'. This yields an enduring connection to a sense of loss and suffering (Levinas 1998) and, crucially, an acute attentiveness (Bloch 1986) to a simultaneous resistance against misery and access *through* misery to revolt (Anderson 2006, p. 701). Through REAs' critical modalities of hope, grief and mourning over the lingering absent presence – and present absence – of cherished and irreplaceable earth kin are interpreted as a call to revolt against the 'Now' and forces that are fuelling the tides of loss.

The common phrase 'at the edge of extinction' (van Dooren 2014; Rose 2017) is an apt one, for many species indeed hover at a precipice – whether they rebound or are lost forever is not yet decided and depends on a host of factors. It's difficult to fathom the utter desolation of a lifeless world, though American science fiction author Alfred Bester offers a painfully convincing depiction of such a hellish scenario in his 1903 eco-apocalyptic short story 'Adam and No Eve'. After a failed experiment sets off an uncontrollable chain reaction of iron disintegrations that scorch the Earth, the only prospects for terra's foreseeable future are as a planet of 'rock and stone, and metal and snow and ice and water' but 'no more life' (Bester in Ashley 2020, p. 313). Yet Bester posits that 'there could never be an end to life' and that 'life would take new root' (p. 314). But in the here and now, terra's incredible life forms are quickly vanishing, and this matters to them,

---

5 'Trace' or *différance* is alterity as such, the differential, self-individuating, prelinguistic (genetic) structural condition of life itself- vegetable, animal, cultural- that precedes the very distinction between *being* and *beings* [emphasis added] (Fritsch et al. 2018, p. 9) and which also remains as exteriority within the 'new' or other.

as it should to us; yet there is still time; so many of these priceless living wonders remain that can yet be cherished and preserved; what once was thought lost can turn out not to be so, as evinced by recent accounts of rediscoveries of species previously believed to be extinct. After centuries of persecution, for example, many whale species had been nearly eradicated from terrestrial existence. Now, some like the humpback whale are making comeback. Absolute numbers of a given population or species don't mean that threats and pressures have lessened or vanished, and a small species population whose decline has stabilised is still at significant risk of extinction (Mehrabi & Naidoo 2022). Nevertheless, such stories demonstrate that what was lost or destroyed seemingly beyond repair can be restored again. Moreover, of the approximately 1.2 million species we know something about, at least 7.5 million more remain yet to be discovered, a testament to the always-unfolding and indeterminate nature of the real and Not-Yet. How do we learn to live well together in and after the 'end', on this damaged earth? How to stem the tides of loss afflicting terra? There's no silver-bullet or permanent solution, for the inescapable politics of cohabitation is a pluralistic and never-ending process laden trials, triumphs and surprises. Nevertheless, below I explore some potential starting points for a terrestrial ecopianism.

Kinship and the Politics of Terrestrial Ecotopianism

> Democracy can only be conceived if it can freely traverse the now-dismantled border between science and politics, in order to add a series of new voices to the discussion, voices that have been inaudible up to now, although their clamour pretended to override all debate: the voices of nonhumans. To limit the discussion to humans, their interests, their subjectivities, and their rights, will appear as strange a few years from now as having denied the right to vote of slaves, poor people, or women.
>
> – Latour (2004, p. 69)

Owing to the decidedly Anthropocentric foundations of Western culture discussed in Chapter 2, it's unsurprising that conceptualisations of politics and political subjects have typically denied humans' ethico-political responsibilities to other terrestrials. In light of the proclaimed insurmountable ontological divide erected between them (Plumwood 2002; Rose 2012), the latter typically don't count as political subjects, whilst humans are posited as 'the centre of the cosmopolitan universe' (Pepper 2016, p. 116). In stark contrast, many indigenous and other cultures see other terrestrials as kin, persons with agency and inherent worth. For instance, Anishinaabe notions of political community and citizenship encompass human as well as other-than-human persons (Borrows 2002). Similarly, the ecotopian manifestations discussed throughout in the (critical) posthumanist tradition (Plumwood 2002; Latour 2004; Woolfe 2010; Rose 2012) variedly avow a non-hierarchically organised 'cosmopolitical interconnectivity' (Pohl 2019, p. 71), and assert that it is our *ethical duty* to live *well* and *justly* with earth others. As alluded to in previous chapters, Both Latour (2004, 2017) and Derrida (2003a, c) variedly critique and deconstruct boundaries in relation to potential ecotopias-to-come and the inter-actant relations that could (and should) constitute them. Ideal societies can never be fully actualised in any present because of the future's irreducible 'otherness' which can never be known in its entirety, and because exclusions – on the basis of race, gender, class, species, etc. – always remain. Steps towards greater inclusivity proceed via always-ongoing processes of deliberation whereby we critically reflect upon, deconstruct and expand the collective 'we' that constitutes the demos in order to include actants that had previously been excluded, and to radically rethink the 'ought' of how we live together. In this vein Derrida's (2003a, 2005c) conception of a 'democracy-to-come' features a striving for a global cosmopolitical alliance that extends beyond the internationality of nation states and traditional notions of citizenship (p. 124). His concept of a 'democracy-to-come' serves the utopian critical function of highlighting the myriad ways in which present 'democratic' modalities remain 'inadequate to democratic demand' – specifically demand by multiplicities of other terrestrials cast outside of the privileged terrain of the human (Derrida 2005c, p. 86).

As we saw in Chapters 3 and 4, Latour's (2004) conception of the collective 'we', firmly rooted in particular places, continuously aims to include previously excluded actants who must be treated as ends in themselves (p. 156). Similarly, the cosmopolitical approach outlined by Latour (2004a, 2004b) and Isabelle Stengers (2010) avows the common construction of a 'democracy of all things' that includes human, other-than-human, and even actants not traditionally considered 'alive'. Boundary delineations – between human and other-than-human dwelling spaces, species, etc. – are deconstructed to varying degrees in the ecotopian manifestations discussed in previous chapters. Yet more often than not, the alternative (and more ethically desirable) modes of relationality that they envision entail critical-posthumanist transitions away from ontologies of separation and epistemologies of domination towards dialogical interactions and partnerships marked by openness, tolerance, empathy and respect for the 'other's difference' (Plumwood 1995, p. 159; 2010, p. 123). Use of others to further one's life and wellbeing is done respectfully, not exploitatively, and others more generally serve as both means as well as ends in themselves. Hence their allusions to more direct-democratic modes of governance, political participation and interpersonal interactions as generally more conducive to the development of more liberated futures (Plumwood 1995). Captain's vision of a world populated by multispecies assemblages, of beings with feathers, some with scales, ad infinitum (Chapter 5) recalls Donna Haraway's (2015) concept of 'the Chthulucene'. The latter points points to how in this era of entangled spatial-temporalities the ethical imperative is to learn to collaborate and live with other terrestrials so as to make possible the flourishing of Earth's rich multispecies assemblages (p. 160). In citing the ethical imperative of making 'kin' with these myriad Earth others in the Chthulucene, Haraway (2015) gestures beyond the somewhat limited understanding of 'kinship' exhibited by some REAs who occasionally prioritise a being's intellectual or social likeness to us, and towards a recognition of the fact that we are all earthlings and thus to varying degrees all kin 'in the deepest sense' (p. 162). Proceeding from this recognition, the task then becomes to carefully and painstakingly chart who we find ourselves dwelling with in particular spatial-temporal localities, and how best to live with them (Latour 2004, 2017).

Derrida's (Derrida & Dufourmantelle 2000, 2005b, 2005d) work on hospitality underscores a key connotation implied in 'conviviality'.

Hospitality in the Derridean sense is no separate realm of ethics but *is* ethics itself manifest spatially as the conditions governing how we live together and relate to one another (Morin 2015). Yet, as explored throughout, hospitality harbours internal tensions and contradictions, for there is exclusion implied in the supposed control and mastery over what one takes to be one's home in the act of extending hospitality. Hospitality as ethics in its unconditional or absolute form is an a priori promise in relation to the 'other' that is not tied to any conditional notions such as right, law, debt, or duty; in effect, it is bound by no laws or limitations (Derrida 2003c; Westmoreland 2008, p. 3). However, 'absolute hospitality' is inherently self-defeating because ethics as systematic code inevitably 'breaks down under the strains of singular responsibility' (Derrida 2003c). What's more, as Chao et al. (2022) observe, 'In ecological communities filled with predators and prey, hosts and parasites – worlds where hostility and hostages are embedded at the very heart of hospitality – alliances can be fleeting as interests align, only to unravel again' (Kirksey & Chao 2022, p. 10). One cannot offer unconditional hospitality to all. Ethics as singular responsibility (in the Levinasian sense) is thus 'impossible because it involves an unconditionality that is untenable' (Attridge, cited in McQuillan 2007, p. 62). Recall in Chapter 4 deliberations amongst community/kin members around limited time and resources, hence their aversion to welcoming outsiders in Saunders' story 'Terranora'. This aporia is what renders infinite and equal ethical responsibility to singular others ultimately 'impossible' in any political instantiation, as the latter necessarily entails a form of calculating responsibility that must take account of various 'others', weigh myriad (and sometimes conflicting) interests and concerns, and ultimately erect boundaries between a collective's inside and outside. Hospitality therefore often manifests as a welcoming of expected guests and hostility towards the unwelcome or unanticipated 'other'. However, neither is complete closure possible, as there is always an unforeseen strange 'other' who arrives (or intrudes upon the self, the familiar, etc.) either by force, surprise, or deception (Nancy & Hanson 2002; Latour 2004), hence the ultimately permeable and provisional nature of all boundaries.

Ethical relationality and responsibility to the 'other' perhaps can never be indiscriminately open; in our necessarily finite interactions with singular others there is always some degree of (violent) exclusion, always someone who is left out (Westmoreland 2008, p. 3). However, the world isn't just filled with predators and prey, but with symbionts too. We ought to strive towards instantiations of the ethical and political which most closely approximate justice. This is precisely what the indigenous sensibilities, REAs and ecotopian texts discussed throughout variedly engage in: deconstructive moves that critically reflect upon ethico-political externalities of hospitality, and more generally the right of existing collective members to decide the fate of those not-yet-arrived. In their deliberations, who/what is taken into account, and whose agency is recognised and respected, includes a great many more beings than has traditionally been the case. Though, like the open-ended utopianism of REAs that never reaches a final end state, the collective never succeeds in taking all into account, nevertheless it should never cease in its attempts to do so. In the spirit of utopianism, the aim is to strive towards and instantiate, as best we can, *better* modes of being, which is a never-ending process. In Le Guin's *Always Coming Home* one encounters a world brimming with purposive agency (see Chapter 4), whose rich cosmological system referred to as the 'Nine Houses of Being' includes:

> the 'five houses of the earth', which includes the earth, moon, rocks, landforms, fresh waters, individual animals, humans currently living, plants used by humans, domestic and ground-living birds, and domestic animals (Le Guin 1985, p. 46); the 'four houses of the sky' include most birds, sea fish, wild animals not hunted for food, plants or animals considered as species or in general, the human race, the dead and unborn, beings in stories or dreams, the oceans, sun, and stars. (Le Guin 1985, pp. 47, 307)

As with many REAs, the abyssal ontological divide (Heidegger 1983, p. 384) between humans and other terrestrials is considerably bridged, and significant strides made towards moving beyond the constitution of other beings through their perceived lack in relation to human Dasein (Plumwood 2002; Elden 2006, p. 280). Even more intriguing, however, is the virtually boundless reach of the Kesh collective which, like many indigenous kinship networks, spans across spatial *and* temporal scales from the past-present to the future-to-come. The Kesh include the departed, the ('to-come') unborn, spectral entities in 'dreams and stories' as members of the broader collective, who are welcomed unconditionally (as integral components of rather than mere arrivants) because one cannot fully anticipate the nature of spectral beings or beings to-come. Elsewhere, the Cartesian subject/object divide (Baker & Morris 2005) appears to re-emerge when certain beings are posited as *either* seen *or* seeing, thinking *or* thought, as with the division between 'wild' and 'domestic' animals into the 'Houses' of the earth and sky. Nevertheless, in the novel other beings are, broadly speaking, not conceptualised antagonistically or as lack in relation to humans, but as equally consequential and agentic (recall Delfin's ruminations on oceanic agency in Chapter 5), who are not merely acted upon or spoken to but who react and respond in turn. Moreover, humans are not the only ones endowed with the right to 'welcome', but also appear as arrivants to be welcomed by other terrestrials: 'I felt light, lying at the side of a small clearing under old bay laurel trees, looking up at the star patterns; I began to float, to belong to the sky. So Coyote let me come into her House' (p. 20). Thus, the novel's human characters are not the only ones to police borders, to decide who is welcomed into a given community and on what basis. In Huxley's *Island,* certain kinds of animals, such as dogs and parrots, are welcomed unconditionally into human spaces of habitation, while the relationship to chickens, for instance, is more ambiguous and qualitatively distinct, designated as they are as *mere* food.

In previous chapters, we saw that often a species' possession of certain characteristics – intelligence, sentience, or ecological function – is typically cited as a justification for its inclusion within the community of ethical subjects. This amounts to a mode of 'ethical' relationality predicated on the logic of 'sameness' that lacks respect or appreciation for the other *as other* (Levinas 1995; Plumwood 1995). Moreover, it suggests the presence of a conditional ethics of hospitality akin to the conditional hospitality of the nation-state wherein the citizen, on the basis of right and law, is defined in at times antagonistic opposition to the foreign 'other' (Westmoreland 2008, p. 2). Often, ecotopian literary and activist sensibilities merely *extend* traditional boundaries around who is welcomed and who matters (i.e. on the basis of cognitive, emotional, etc. capabilities, ecological significance), but boundaries remain and are implicitly if not explicitly 'policed'. For those of a biocentric (life-centred) ethical stance, the boundary extends all the way to that which is alive, therefore more closely approximating unconditional ethical modalities, even though this approach also involves establishing yet another untenable binary between life and nonlife. Hence why an ecocentric sensibility as espoused by many indigenous peoples is preferable in that there is no centre, no fixed hierarchy or rigid binary dividing those who matter from those who don't; rather, value and personhood are diffuse throughout natural assemblages and their myriad constitutive agents. But as distributive justice, and ethics as such, cannot be absolutely or universally applied, perhaps priority should be given to actants exhibiting a capacity for flourishing, with an awareness of the considerable epistemological and methodological difficulties of assessing such capabilities. Negotiating ethical modes of relationality with hurricanes, for instance, is considerably more complex and perhaps ultimately unfeasible, though such entities should feature in any negotiations of the coming collectives. Furthermore, unlike with conditional modes of hospitality accompanied by laws articulated via a shared language, ethical encounters with animal 'others' (and between humans and other animals more generally in some of the ecotopian texts) don't necessitate linguistic exchanges as a condition for hospitality. Thus, a central component of the standard conditionality of hospitality as ethics is absent in many of the ecotopian manifestations discussed in this book. This gestures towards a promising (ecotopian) way forward: striving to welcome irreducible 'others' on the basis of our shared mortality and co-evolutionary kindship bonds.

As discussed in Chapter 6, even kinship serves as a two-sided ethical coin which includes as well as excludes (Pyle in Whitehead 2020, p. 87). Similarly, the very act of 'welcoming' can harbour violent exclusions, wherein it is presumed that 'we' – whether humanity alone or a more expansive collective – are the rightful inhabitants at home in a place, and who can thus treat the 'arrivant' as an always-temporary 'guest' who must bend to our terms of cohabitation (Derrida 1999b; van Dooren 2014). In this sense, Latour's (2004) conceptualisation of the cosmopolitical collective-to-come perhaps affords a little too much primacy to the existing collective's right to decide, pending compromise and accommodation, who is to be 'welcomed' and on what terms. For Latour the latter is predicated upon the degree to which new arrivals can harmoniously mesh with existing collectives by finding their rightful place in it, and on *condition* that they don't fundamentally disrupt the already-existing order (pp. 107–110; Morin 2015). The query, 'Can we live together?', is posited as the sacred duty of those in the already-established collective rather than something that might be equally posed by those on the outside; thus, the collective's perimeter, however tentative, is still policed by those on the inside who ultimately hold the power to decide on affairs (Morin 2015, p. 37). For instance, the Crystal Waters ecovillage in Queensland, Australia is a lived ecotopia covering 640 acres of bushland and founded on permaculture principles (Crystal Waters 2022). The community of over 240 human denizens and their residential and commercial spaces occupy a comparatively small portion of the land, with the rest designated 'for wildlife'. Yet this 'peaceful' cohabitation is founded on an exclusionary 'No cats & dogs' policy in order to reduce the risks that predation to local wildlife. Perhaps such exclusions are unavoidable in the business of building and maintaining collectives. To reiterate, the politics of living together is never devoid of conflict, and the process of expanding and/or reconfiguring boundaries is always ongoing; yet remaining critically reflexive of these implicit power dynamics and their potential ethico-political ramifications is essential. Moreover, determining who is host, guest, inside or outside of a collective, as with rigid dichotomies between 'self-identity' and 'absolute alterity' is not only ethically problematic but also ontologically dubious. Like the self, the perimeter of the collective is always porous, unstable and in constant flux.

Thus, it's crucial to remain critically aware of the potential undemocratic and unethical implications of exclusions resulting from collective boundary delineations (Watson 2014). When Latour ruminates on what obliges one to 'reserve the water of the river Drome for fish as opposed to using it to irrigate corn fields subsidized by Europe', the answer doesn't merely lie in whether or not we've taken into account all entities affected by such an act (2004, p. 198) or in considering how excluding fish, for example, will affect the existing collective. Latour (2004) notes that in such instances morality should change direction, from residing in foundational ethical precepts to lying ahead in a continuous process of registering the 'appeal of excluded entities' (p. 198). Yet, I would venture to posit that something is lost here that is essential to truly ethical modes of relationality

(Plumwood 1995; Derrida 2003c, 2005b, 2005c): consideration for the singular, irreducible 'other' *as such*, as well as for that/whom which might arrive without prior notice and therefore might surprise us. It is not enough merely to 'register' the 'appeal of excluded entities'. Moreover, depriving the other (fish) of water, or excluding any appellant for that matter, is ethically problematic because the *fish* needs water in order to survive. This is precisely the message that REAs and many indigenous cultures, pulled as they are by the ethical appeal of the other, strive to bring to the forefront: that more ethical modes of relationality require that we attend to the wellbeing and perspectives of the other *as such* and *on their own terms*, rather than on the basis of ultimately self-regarding concerns over whether our treatment of them might negatively impact *us*. Latour's (2004) cosmopolitics (indeed any articulation of the better), though gesturing towards more pluralistic conceptions of the 'we', would benefit from a more utopian openness to the heterogenous and unforeseen other, whose arrival, as the unanticipated excess that disrupts the 'Now'/self-same, harbours vital transformative potential. Moreover, Latourian (2004) cosmopolitics perhaps might benefit from being supplemented with more nuanced ethical considerations such as those that arise from Plumwood's (2003) theory of ecological animalism (see Chapter 4), which urges that we treat others as *with* and not *for* us (Le Guin 1987). Similarly, a Derridean overemphasis on the unbridgeable singularity of each being risks overshadowing the dynamic interrelationality that comes prior to the singular self and constitutes our shared terrestrial existence.

Towards Respectful Conviviality

In sum, the politics and ethics of living together – as with utopian theory and praxis itself – is inescapably complex, ongoing, and laden with trials as well as wonders. Yet, live together we must, ideally in ways that considerably minimise wanton and systematic destruction of our terrestrial support systems. What might 'ecotopian' modalities of cohabitation on a damaged planet look like? How do we begin to build spaces more conducive to multispecies flourishing? Drawing on Latour and Stengers' cosmopolitical approach, Narayanan and Bindumandhav's (2019) concept of 'posthuman cosmopolitanism' seeks to inform the development of 'species-inclusive cities or zoöpolises' (Seymour & Wolch 2009) by promoting an ontological vision of cities as multispecies assemblages involving different forms of worlding, and thus requiring attentive care and constant negotiation. Recalling Solarpunk's visions of multispecies urban cohabitation in *Multispecies Cities* and similar examples from the literary ecotopias surveyed in Chapter 4, promising developments can be seen in biophilic and hybrid nature-culture urban design initiatives around the world (McDonald & Beatley 2021). Biophilic design draws on the biophilia ('love of life') hypothesis popularised by socio-biologist E. O. Wilson (1984), which posits that, due to our extensive co-evolutionary histories and entanglements with other life forms, humans have an innate desire to affiliate and commune with them. To varying degrees, these innovative experiments in urban redesign for multispecies cohabitation seek to dismantle human/nature binaries by 'bringing nature into cities' (Owens & Wolch 2019). Examples include expanding 'green infrastructure' in urban spaces in the form of parks, rooftop gardens and 'greening' the facades of buildings, and building wildlife corridors to increase habitat connectivity for enhancing biodiversity. Alan Marshall's work on ecotopian cities (2015) has been instrumental in shedding light on the boundless potential of such developments. An abundance of evidence suggests that biophilic urban design confers a number of benefits for humans – i.e. reduced violent crime rates,[6] improved mental health, improved air quality, smaller carbon footprints and improved resilience to climate shocks (Yin et al. 2018). However, a greater focus is needed on how we might design these shared spaces so that they're conducive to the wellbeing of our terrestrial kin as well as our own. Another promising development is Büscher and Fletcher's (2019, 2020) 'convivial conservation' approach that challenges capitalist expansionism and entrenched human/nature dichotomies in mainstream conservation, and seeks to

---

[6] Of course, this is no substitute for addressing some of the structural drivers of violent crime such as socioeconomic inequality!

explore ways in which humans might live well *with* 'nature' (9). Büscher and Fletcher (2019) call for a discursive shift in conservation from protecting 'nature' *from* humans (which effectively rehashes Nature/culture dualisms) towards promoting nature 'for, to and by humans' (2020, p. 163). Although, their approach does remain quite anthropocentric in its emphasis on nature for *humans*, whereas we ought to be preserving and protecting natural 'others' for their own sake as well. Nevertheless, the convivial approach to conservation (Büscher & Fletcher 2020) harbours considerable ecotopian potential in its attempt to promote integrated spaces for multispecies cohabitation, localised and democratic forms of governance. Convivial conservation also does well to draw attention to the role of power within biodiversity politics and conservation, arguing that conservation initiatives should target upper-class elites and the structures and relations of capitalist accumulation that they reproduce, rather than vulnerable local communities. As such, the approach urges widescale repossessions and redistributions of land and resources to the communities from whom they were stolen, and actively involving local communities as co-managers in conservation initiatives (Büscher & Fletcher 2019).

As discussed throughout this book, living well with our terrestrial counterparts will necessitate fundamental changes in attitudes, values, worldviews and sensibilities: in effect, in how we view, value and interact with our terrestrial kin. It will require an ethical (re)orientation wherein, like many indigenous cultures, we embed care and respect into every aspect of our dealings with others. Among other things this would require a breakdown of hierarchical and dualistic conceptions of human relations with the wider world, becoming intimately attuned to the 'nature' all around, cultivating the art of listening and developing a heightened awareness of the distinct sights, sounds and rhythms of local flora and fauna, and overall being more 'alive' to ...

> the touch and scent of the air on the cheek, and the fall and motion of the light across the air, and the colour of the grass on the high hill across the river, and the thoughts of the body ... the shimmer and ripples of colours and sounds in the clear darkness of the depths, endlessly moving, endlessly changing, endlessly new. (Le Guin 2012, p. 201)

This will be far from an effortless or uncomplicated task in many cases. As Rose Bird (2017) points out in relation to human-monk seal relations in Hawaii, the effect of different histories, colonial legacies, and knowledges is that 'some people recognise monk seals as family, while others claim that they are an invasive species, and many subsistence fishermen claim that monk seals steal their fish' (p. 130). Of course, fishermen must make a living during these precarious times; however, fish belong neither to them nor anyone else. Moreover, Monk seals' arrival on the Hawaiian islands predates that of humans, so if anything it is the latter who might more rightfully be considered 'invasive' (Rose 2017)! During times of widespread socio-ecological upheavals, hybrid and shifting nature-culture entanglements, the very notion of 'invasiveness' or 'alien' is rendered additionally dubious. What's more, migrating in order to flee threats and in search of food, safety, and habitable dwelling spaces is the right of all terrestrials; movement has been an essential feature of life since it made its first stirrings around 3.5 billion years ago (Mancuso 2021). During the climate-ravaged decades ahead, movement will be an indispensable survival strategy, which should be denied to none. Learning to live alongside potentially dangerous species, such as venomous snakes and large carnivores add further layers of complexity and challenges. However, rather than immediately resorting to lethal methods in conflict situations, non-lethal methods for maintaining safe distances and minimising conflict[7] – i.e. translocation or the use of 'natural barriers' in the form of fences combined with fast-growing trees and plants (Nyhus 2016, p. 156) – can be sought. The (ecologically successful) reintroduction of grey wolves into Yellowstone National Park (US) in 1995 is a commonly cited example of tensions resulting from cohabitation with large carnivores, who often pose added risks to humans and their 'property' – i.e. livestock (Nyhus 2016). In this and similar cases, a different biocultural setting with socioeconomic support for farmers whose livelihoods are threatened by wolf predation, along with a newfound respect for wolves and their needs, might yield different results. Restoring wolves' natural prey

---

[7] However, as Narayanan and Bindumandhav (2019) emphasise, the role of religious superstitions in moulding perceptions of certain nonhuman animals like snakes adds further complexity to such cosmopolitical endeavours.

populations would further help minimise incidences of conflict with humans (Kuijper et al. 2019). In major cities like Los Angeles and Mumbai, humans are learning to live alongside big cats (The Guardian 2022). Greater understanding of the movements and habits of these animals has helped inform policies for wildlife-friendly infrastructure (i.e. wildlife crossings) (Smith et al. 2017). Notably, in Mumbai, one of the most densely populated cities on earth, education programmes designed to help the cities' human denizens cohabit peacefully with leopards – for instance, by maintaining respectful distance, going out in groups rather than alone at night – have greatly reduced the risk of accidents and conflicts.

Toncheva et al.'s (2021) fascinating study of human-bear relations in Yagodina, a village in the Rodopi mountains of Bulgaria, through a convivial conservation lens sheds intriguing light on the (ecotopian) possibilities of cohabitation. In Yagodina, agricultural decline which left vast spaces of previously cultivated land free for an unplanned rewilding, in addition to bears' newly protected status, saw local bear populations increase. Additionally, biodiverse-rich forests surrounding the area provide plenty of resources for bears and humans to thrive, thereby reducing the risk of conflict and helping to ensure successful cohabitation (p. 5). Crucially, cohabitation is:

> determined by the attitude of *both* [emphasis added] species who attempt to avoid one another and do not enter conflict situations. Respondents described various methods that could potentially diminish encounters: making a noise, avoiding areas known as bear habitats ... (p. 5)

Cohabitation, in other words, involves ongoing, careful and respectful negotiation by parties involved around what each need in order to live well and flourish. A mutual flourishing is the stated aim of environmental NGOs like Rewilding Europe, which has overseen several rewilding projects across Europe including the reintroduction of Bison in Kent, England in July 2022 (Boyd 2022), although the organisation's emphasis on the 'business' benefits of rewilding somewhat misses the mark. Nevertheless, as discussed throughout, cohabitation doesn't mean the absence of conflict or predation, or that there shouldn't be some boundaries in place for facilitating respectful coexistence; Toncheva et al. (2021) emphasise that in Yagodina, a common division of space persists between the humans' domain in the village, and the bears' in the forest. Though these spaces somewhat overlap, as there are no PAs in the area and thus no strict lines demarcating 'human' and 'bear' domains, each should be respected by the other party. This has allowed the humans and bears of Yagodina to negotiate their interactions in a way that's sensitive to the situated needs of this particular community, rather than according to externally imposed understandings of what conservation should entail – i.e. strict separation, managing the behaviour of species in question. Spaces where both species interact include rivers and meadows. Villagers navigate these shared spaces with care and acute attention to signs indicating the bears' presence, such as overturned stones. This fluid, context-specific approach to conviviality has, as the authors observe, allowed the human and bear residents of the region to effectively 'learn how to live together' (p. 9). The authors highlight two crucial factors contributing to the relative success of this cohabitation story: (1) the villagers' non-dualistic views of human-nature relations, wherein as noted above only the village space is designated as a 'human' domain, and relatedly (2) the villagers' intimate knowledge and consideration of the bears' behavioural patterns and needs. The bears are seen as 'fellow inhabitants' by the villagers occupying a shared space. Once again, respect, attentiveness and care thus re-emerge as key preconditions for successful cohabitation. As Kimmerer's (2013b) honourable harvest attests to, living well with others requires that we don't take too much – food and other resources, space – so that others may have enough in order to thrive.

Sharks have roamed the seas for over 400 million years, and survived the previous five mass extinction events on earth. I had the indescribable joy and honour of interacting with scores of them, from tiger sharks to bull sharks, whilst interning for the R. J. Dunlap Marine Conservation Program at the University of Miami as a final-year undergraduate student in 2014. Now due primarily to overfishing, climate change, pollution and related pressures, the global abundance of sharks and their ray relatives has declined by a staggering 71% since 1970 (Pacoureau et al. 2021). They also happen to be central villains in much of the Western cultural imaginary, a widespread fear and hatred that is expressed in and exacerbated by films such as the Jaws franchise, The Shallows, Deep Blue Sea, and countless others. Yet, in addition to the ethical imperative of learning to live well with our marine counterparts, sharks are keystone species that play a vital role in maintaining the integrity of marine ecosystems. Our wellbeing and continuity are inextricably linked to theirs. Here too, a (re)orientation in sensibilities and values is of the essence, difficult though the above sentiments may be to uproot. For the Kanaka Maoli (indigenous Hawaiians) briefly referenced in Chapter 6, sharks (Mano) are no bloodthirsty monsters but culturally significant individuals with unique personality traits (i.e. guardians, angry, provoked). Those with whom native Hawai'ians developed intimate relations even had their own personal names (Puniwai 2021). Punaiwai (2021) cites the traditional saying '*Pua ka wiliwili, nanahu ka mano*' (when the wiliwili blossoms, the sharks bite) (Puniwai 2021, p. 12). Like the villagers of Yagodina, the Kanaka Maoli have cultivated an intimate awareness of the flora and fauna persons in their communities through prolonged experience and acute attentiveness to their lifeways. As a result, they came to note a wider ecological nexus connecting marine and terrestrial processes; through careful observation, they realised that the blooming of the wiliwili flower happens to coincide with the mating season of local sharks, hence their heightened arousal and increased likelihood to bite during this time (Puniwai 2021). As the author of the study observes, such deep observation can offer important lessons for how to minimise human-shark conflict and generally coexist more harmoniously with them. Care and attention take time and resources that not all have access to – hence why this work must go hand in hand with radical reductions in socioeconomic inequalities between humans. However, telling new stories, restoring marginalised ethical practices and 'actively cultivating relationships' (Puniwai 2021, p. 16) that foreground care, attentiveness, respect and reciprocity can help us begin the process of restoring relations with our terrestrial kin.

Ecotopias in the Mainstream? Iterations of the Green New Deal

> The dramatic crises currently afflicting humankind and the planet require responses of a very different *quality* than those offered by governments and the UN system. They require an *alternative civilisation paradigm*.
>
> – People's Summit Rio+20 (2012, p. 8; emphases added)

Bloch's framing of utopianism as transcendent without transcendence refers to the notion that all utopias are necessarily and variedly limited by the material confines of the present. However, in casting their sights to other ways of being, they negate and transgress their the limitations of their deficient 'present'. The precarious present demands radical – indeed, utopian – transformations in late capitalist societies. Disillusioned with continued technocratic responses to mounting social and ecological decline, the Thematic Groups at the Thematic Social Forum of Porto Alegre (TSF 2012) produced the decidedly utopian working paper *'Another World is Possible'* at the alternative People's Summit Rio+20 in the summer of 2012. The text is infused by Buen Vivir (or 'to live well'), a 'heterogenous' conjunction of alternative views and approaches to development and deliberations surrounding the 'good life' by indigenous movements and critical scholars across Latin America (especially Ecuador and Bolivia) (Gudynas 2009, p. 50). Buen vivir draws on the Quechuan concept *Sumak kawsay*, wherein Sumak denotes that which is full of plenitude, is sublime, excellent, magnificent, beautiful; Kawsay refers to life as dynamic, active existence. From the perspective of Buen vivir, the 'good life' can only take place in community with other persons and nature (Gudynas 2011, p. 442). As such, the paper delivers fundamental critiques of the Western-Capitalist development paradigm, namely its unquestioned anthropocentrism and reliance on an exploitative and expansionary economic model. We thus face an urgent dilemma: 'continue down the path of production, depredation and death, or embark on the path to a new model of sustainable civilization, respectful of life and reconciled to nature' (2012, p. 29). The first and foremost step, they urge, is philosophical: we must change how we view ourselves in relation to other terrestrials and the wider ecological constellations that constitute life. Echoing indigenous exhortations across the globe, the paper calls for a shift towards an ontology of interconnectedness wherein humans are re-situated on terra in inextricable relatedness with other all other persons. It urges the realisation of 'Mother earth' as a living entity composed of innumerable beings, with certain inalienable rights.[8] This would also entail an ethical shift in how we live with and interact with others. Though we must consume and take life in order to ensure our lives, such patterns of production and consumption 'cannot be of unlimited destructive development at the cost of other peoples' (2012, p. 30). The rights to life of anyone being must ultimately be limited by the rights of others (2012, p. 30).

Proposals for a 'Green New Deal' (GND) have since proliferated across the globe as governments have, at least seemingly, begun to take notice of mounting alarm bells over contemporary socio-ecological breakdown. The GND was initially introduced in 2019 by US senator Ed Markey and congresswoman Alexandra Ocasio-Cortez as a decidedly progressive manifesto for helping the US reach carbon neutrality by 2030 by, for instance, creating millions of 'green' jobs, expanding access to affordable healthcare, and initiating significant infrastructural and manufacturing transformations for sustainability (Nilsen 2019). The various subsequent iterations of the 'Green New Deal' in one sense harbour a utopian impulse in that they aim to challenge and move beyond the most severe socio-ecological depredations of unbridled capitalism. Thus, versions variedly focus on the need for 'just' transitions towards carbon neutrality (Zografos & Robbins 2020). These rather bold initiatives involve some aspects of denunciation and annunciation – i.e. the gradual phasing out of fossil capitalism, reducing extreme socioeconomic inequality, promoting racial and gender equality and general socio-ecological wellbeing – both of which are essential if utopianism is to effectively and justly transgress the confines of the present.

---

8   i.e. 'the right to life and to exist, to be respected, to regeneration and biocapacity, the continuation of cycles and vital processes, to maintain their identity and integrity and integrity as different beings, self-regulated and interrelated', as well as rights to water, clean air, health, etc. (2012, p. 30).

The UK Student Climate Network's vision of a UK GND revolves around five key principles: (1) complete decarbonisation of the UK economy in a way that reduces extant socioeconomic and other forms of inequality, (2) the creation of millions of new 'green' jobs, (3) making the financial system democratically accountable, (4) the restoration of ecosystems and vital habitats for all, and (5) helping others decarbonise justly (UKSCN 2022). Similarly, the EU has laid out its vision for a 'European Green New Deal' in its 'Fit for 55': delivering the EU's 2030 Climate Target on the way to climate neutrality. Amongst other things, it aims to transform the EU into a 'modern, resource-efficient and competitive economy' by ensuring that there are no net GHG emissions by 2050, that economic growth is decoupled from resource use and that such growth is just – that is, leaving no one behind, in the spirit of the wider Sustainable Development framework (ECCEU 2022).

However, for one thing the idea of decoupling economic growth from resource use and ecological degradation via technological innovation, otherwise known as 'green growth' or 'ecological modernisation', is a contentious one that remains unsupported by empirical evidence (Hickel & Kallis 2020). A parallel development is the process of 'greenwashing', when corporations co-opt sustainability narratives and tout their purportedly green credentials for boosting profits and credibility, whilst rarely making any fundamental changes to their practices. More problematically, and somewhat ironically, Green New Deal proposals, especially in their EU and other Global North iterations, risk replicating 'climate colonialism' and the proliferation of 'green sacrifice zones' (Zografos & Robbins 2020). In an excellent 2021 article on how seemingly 'sustainable' solutions often come at a deadly price elsewhere, George Monbiot (2021) uses the example of the electric car to highlight how the push for cleaner vehicles is leaving a litany of social and ecological disasters in its wake. The photovoltaic cells and batteries that form the essential building blocks of the 'green energy and transport revolution' are 'being built with the help of blood cobalt, blood lithium and blood copper' (Monbiot 2021). The global race to reach Net Zero by 2050 has fuelled a surging demand for rare earth metals such as lithium, nickel and cobalt. Cobalt, the modern-day 'oil' of a low-carbon economy (Sovacool 2019, p. 915), is a critical mineral used to manufacture phones, computers, electric vehicles, wind turbines, lighting, solar panels, fuel cells and nuclear reactors (Herrington 2021). These non-renewable raw materials primarily sourced through mining are often associated with among the highest rates and risks of deforestation, land grabs and similar socio-ecological injustices in places like the Democratic Republic of Congo, which currently provides over 60% of the global supply of cobalt. Some estimate that meeting the goals of the Paris Agreement will require a fifty-fold increase in electric vehicle adoption between 2016 and 2030; as such, demand for cobalt, one of the essential minerals used in the construction of lithium-ion batteries for electric vehicles, is expected to increase 500% by 2025 (ERG 2018; Zografos & Robbins 2020). One would shudder to fathom, though can likely anticipate, what further pressures will be placed on ecosystems containing substantial reserves of these prized minerals, and on their inhabitants.

A mad push for individual vehicles, electric or not, as has been the tendency amongst mainstream approaches to energy and transport transitions, is not the way forward. This is characteristic of neoliberal capitalism's individualisation of collective problems with structural underpinnings. Wherever possible we should be pushing not for new 'green' individual cars but for affordable and widely accessible public transport, bicycle use and other low-impact forms of transport, and, crucially, reducing the need to travel long distances in the first place. The green growth or 'ecological modernisation' approach so far touted by transnational corporations, supra-national organisations and governments thus amounts to little more than a thinly veiled continuation of the same. As degrowth economist Serge Latouche (2009) reminds us, it's not enough merely to attempt to internalise the social and ecological costs of growth; we are in dire need of real alternatives which treat none as expendable or as mere resources for others' benefits. Alternatives, that is, such as those proposed by the decidedly utopian 'degrowth' movement of scholars and activists:

> It seems intuitive that if as a society we are to stay within ecological limits we will have to do with less high-speed transport infrastructures, space missions for tourists, new airports, or factories producing unnecessary gadgets, faster cars or better televisions. We may still need more renewable energy infrastructures, better social (education, and health) services, more public squares or theatres, and localised organic

food production and retailing centres. We need therefore a 'selective degrowth'. (Kallis 2011, p. 875)

When discussing any ecotopian project or proposal, therefore, we must continually ask who benefits and who potentially bears the costs? Who might be 'left out'? In this vein, Max Ajl's *A People's Green New Deal* (2021a, b) casts a critical light on the Eurocentric nature of mainstream GND proposals, arguing for their active decolonisation, respect for national sovereignty and reparations for ecological debts accrued from colonial and settler-colonial accumulation on a global scale which 'occurred alongside unequal access to waste sinks and the biosphere's capacity to absorb and process all manner of waste, especially carbon dioxide' (2021b, p. 380). Among other things, Ajl calls for the 'decolonisation' of the atmosphere in the way of, for instance, radical reductions in emissions on behalf of the worst polluters, and a fairer distribution of atmospheric space so that long-struggling nations might find some space to flourish within the wider 'equillibrium of Mother Earth' (2021b, p. 381). Building on this, along with others (Zografos & Robbins 2020) I concur that GND proposals hold some (albeit presently limited) utopian potential. However, I would argue that GND proposals could be made 'utopian' by incorporating explicitly anti-capitalist, decolonial and degrowth perspectives for more concretely diagnosing the ills of the present system and as guides for more just and inclusive pathways. Something, for instance, along the lines of the aims of 'Pacto Eco Social del Sur', a collective of individuals and organisations across Latin America calling for transformative socio-ecological changes for a post-capitalist and post-extractive future that prioritises life over profit-driven accumulation (Pacto Ecosocial del Sur 2022). Similarly, the Indigenous Environmental Network's proposal for 'just transitions', which wholeheartedly rejects the 'commodification of nature' through green capitalist measures like carbon markets, and instead demands legal recognition of nature as a 'rights-bearing entity', respect for indigenous land sovereignty, and an array of community-based projects for multispecies restoration (IEN 2023).

Following on from this, GND proposals often have a glaring omission at their core: our terrestrial kin. Thus, they would do well to incorporate an emphasis on multispecies justice rather than mere 'climate justice' which, as discussed in Chapter 3, remains all too human-centred. Indeed, the European GND features a biodiversity conservation strategy for 2030 which includes a new 'Nature Restoration Law' aiming to 'restore ecosystems for people, the climate and the planet' (European Commission 2022). Among other things the strategy seeks to increase the amount of urban green spaces, stem the decline of pollinators, and related measures for restoring marine and terrestrial habitats across the EU by 2050. However, none of the proposals I've yet come across, including Ajl's notable postcolonial refinement in the form of the 'people's GND', feature an explicit focus on other species or terra as inherently valuable entities who matter in their own right, and who ought to be central protagonists in the transition to more sustainable futures. The emphasis, yet again, remains largely on restoring biodiversity and ecosystems that we deem most essential for ensuring our own long-term continuity. Mining for the materials that we need to fuel the 'green revolution' of course doesn't just generate harmful effects for vulnerable human communities at the front lines of these extractive enterprises, but also for other terrestrials, who are often negatively impacted at a multitude of spatial scales through land degradation, toxification (Sonter et al. 2018) and related 'externalities' of the 'green revolution'. Ecotopian politics would thus require carefully weighing and negotiating these risks alongside the long-term benefits conferred by renewable energy technologies for transitioning towards carbon-neutral futures. Crucially, setting thoroughly just and inclusive transitions in motion requires that we do away with the very notion of sacrificial peoples and places. The world is round – there are no externalities, nowhere to safely or permanently lock away the consequences of our actions. An intervention 'here' always generates feedback mechanisms, foreseen and unforeseen, over 'yonder'. And it is our ethical responsibility to register and listen to the protestations of those variedly affected by us as we make our way forward.

CHAPTER 8

Towards Multispecies Flourishing

> I love you. And even on your sickest day, you deserve an ocean as blue as your name. You deserve a safety as deep as your need. You deserve food, community, school and home ... Know that I love you in a world that would turn your blood into candy. Criminalize your flexibility. Oversimplify your claim to your own life. Know that for me your blood is a scripture. I want it to stay in your veins. I am saying your name. We're not breathing in vain.
>
> – Gumbs (2021, pp. 40, 43)

In her timely monograph *Green Utopias: Environmental Hope Before and After Nature* (2018), green utopian scholar Lisa Garforth urges that the socio-ecological perturbations of the present require a 'green utopianism' predicated on radically new way of being and inhabiting the world, and 'attuned to the complex ways in which we are simultaneously matter and culture' ... 'we need new ontologies, new ethics and new ways of thinking about better greener worlds' (pp. 4, 5). This book has been an attempt at tentatively sketching the contours of – and advocating for – a new green utopianism fit for actively resisting the multiple deficiencies of the present, and crucially, charting more liveable pathways for terrestrials to live better together in the damaged and uncertain terrain ahead. I refer to it as terrestrial ecotopianism, traces and glimmers of which can be seen in indigenous ethics, politics, onto-epistemological systems and speculative fiction, as well as in many contemporary radical ecotopian movements and literary texts. Terrestrial ecotopias build on critical and grounded utopian modalities which emphasise the 'Nowness' of the utopian impulse, with a focus on the here below instead of attempting to beam into the 'beyond' of distant times and places. The sense of the 'beyond' suggested in this book's title is a reference to what might yet transpire in the vast ocean of potentiality that always lies before us. There's a world of difference between terrestrial ecotopias and techno-capitalist 'post-terrestrial' utopias (Alberro 2022) which seek to abandon terra, earth kin and our responsibilities to them in search of new colonial frontiers on other planets. Goodman et al.'s notable 2014 article 'Mars can wait: Facing the challenges of our civilisation' about the criminal waste of resources pursuing cosmic fantasies that should instead be used to ameliorate a litany of human ills and pathologies here on earth can be applied to the human-Earth interface amidst climate and ecological breakdown. Whilst billions in funding continue to be poured into the new space race,[1] vulnerable regions continue to drown, burn, and starve. Rather, we need every resource, every tool in the shed to salvage what we still have here on terra. Hence why terrestrial ecotopias are and must be intersectional, recognising that oppressive cultural dominants have similarly exploited and devalued human as well as other-than-human peoples.

Terrestrial ecotopias further exhibit a healthy degree of open-endedness and critical self-reflection– particularly around boundaries and their exclusions (i.e. self/other, Nature/culture, ecotopian collectives); relatedly, they often place particular emphasis on the process of utopian resistance and transformation, on critical engagement with and transgression of the imposed limitations of the present (Harvey 2000), rather than fixed end-states, yet without eliding the ever-unfolding not-yet. Crucially, they underscore the need to face up to our responsibilities to other terrestrials and resist further decline by (re)creating more habitable spaces in the here below. A key aspect of this work involves resisting and working to 'end' capitalism as one major source of interspecies exploitation and ruination (Gumbs 2021).

Terrestrial ecotopias are firmly rooted in terra and in dynamic and inextricable entanglement with agentic earth others who precede, constitute, always exceed and often surprise us. Of course, a focus on terrestrial groundedness and relationality does not preclude orientations to the

1 Apparently the 'New Horizons' rocket used to explore Pluto for the first time in 2015 cost $720 million (Prescott & Logan 2017). Of course, as the authors point out, this is but a miniscule fraction of the annual US defence budget. Nevertheless, the authors ponder why 'such (publicly funded) Apollo-like "moon shot" efforts are not directed at grotesque problems here on Earth, first' (Prescott & Logan 2017, p. 22).

beyond whether spatially or temporally; movement to new habitable places might entail extra-terrestrial ventures at some point. However, proceeding with the same colonial imaginaries that have led to socio-ecological ruination here on terra is a recipe for further catastrophe. Rather, an ethico-political orientation predicated on respect for and responsibility to our kin, and gratitude for the multiple agencies who sustain us and make 'us' possible, can help break such harmful cycles. Terrestrial ecotopias thus variedly draw on critical-posthuman tributaries, and on a rich tradition of indigenous cosmologies for whom our co-terrestrials are no mere peripheral problematic or resource for ensuring human continuity. Our terrestrial kin – in all of their irrecuperable singularity – occupy the foreground of earthly happenings alongside humans and emerge as central concerns within these manifestations. We are no longer the central protagonists of terrestrial stories. Moreover, terrestrial utopias not only widen but critically reflect on and deconstruct spheres delineating who matters when deliberating on the nature of ideal ecotopian collectives. Their utopian 'ought' and conceptualisations of the good life actively foreground the existence, needs and concerns of other terrestrials through a palpable undercurrent of a striving for multispecies justice. Challenging though it may be to understand and approximate the unique needs, interests and vulnerabilities of our co-terrestrials, we might make a start by cultivating a hyper-sensitivity and awareness to the multifarious narrative agencies that continually surround us. We would also do well to declare all creatures our fellows (Derrida 2009), as sacred entities who are beyond price and thus 'cannot be bought or sold' (Starhawk 1993, p. 18). Recalling Huxley's *Island*, education programmes designed to 'curb the contemporary disconnect from nature' particularly from early childhood (Anderson 2010; Prescott & Logan 2017, p. 12) offer one amongst numerous helpful mechanisms (along with radical reductions in intra and international socioeconomic inequality) for developing the humility, care and respect that are essential foundations for doing better in our dealings with earth others.

Terrestrial ecotopias value diversity and the others' irreducible singularity whilst emphasising the fundamental 'being-in-common' that characterises life on terra (Plumwood 2002). Their utopian impulse draws on and mobilises an ethics of care (Rose 2012; Chatzidakis et al. 2020) and politics of love (Hardt & Negri 2005) that encourages one to not just look at 'but really *see* the warblers, waters and walnuts around us' (Anderson 2010, p. 180). This requires cultivating what Deborah Bird Rose (2013) terms 'active attentiveness' to the agencies around us, being present and alive to their habits and movements, needs, vulnerabilities and desires. However, at its most basic level, love for other species entails ceasing to systematically destroy and exterminate them (McKenna in Ashley 2020, p. 277); it enjoins us to hold everything dear (Berger 2008) in our dealings with other terrestrials, particularly amidst times of systematic and widespread loss which render especially precious every singular moment, individual, and co-evolutionary 'knot' in time – each a unique constellation of memories, experiences, aptitudes, fears, joys, desires that cannot be replicated or recouped once lost. So, Berger (2008) urges us to cherish every wingbeat, every song, every co-terrestrial with whom we share this finite earthly mortality ... in effect, we should 'be careful of each other, we should be kind while there is still time' (Larkin in Muldoon 1998, p. 158). Through these changes in sensibilities, worldviews and ethical modalities we can begin to work towards a more respectful conviviality conducive to a mutual flourishing. With sufficient time, resources and effort, we can become more attuned to local ecological cycles and patterns, and come to learn what resources, spaces, and the like are important to species we co-habit with, as well as what adversely affects their wellbeing.

The other sense in which the emphasis on time is important is because adequate care and attentiveness require time and resources. The latter are increasingly in short supply with many frantically trying to keep pace with the mad treadmills of work, production and consumption that characterise global neoliberal capitalism, with creeping casualisation, mounting pressures and expectations without consequent increases in pay and support. Increasingly there is less time to think, to situate and ground oneself. Hence why we also need radical socioeconomic, political and cultural changes, towards systems that value slowing down and taking the time to take notice of the qualitative aspects of 'the good life'. Then we can begin to understand, for instance, the ways in which noise pollution harms cetaceans and other marine kin, and therefore reduce or altogether eliminate its sources; knowing that high-speed watercraft often prove fatal to manatees, we can reduce the speed of such vessels or even ban their use in manatee territories (Laist et al. 2006). Rather than opting for perfectly manicured landscaping in service of human aesthetic preferences, we'd preserve and promote hedgerow heterogeneity since they provide vital sources of food, shelter and breeding sites for countless species in otherwise denuded landscapes across the UK, Western Europe, and elsewhere (Graham et al. 2018). We'd turn down the lights or opt for long-wavelength lighting in residential and commercial buildings alongside beaches where sea turtles return to nest, so that hatchlings, spurred by their ancient drive to follow the moonlight towards the sea, aren't lead astray to their peril. In developing renewable energy technologies for powering low-carbon and carbon-neutral lifeways, for instance, we'd paint the rotor blades of wind turbines black in order to reduce chances of collision fatalities amongst our avian kin (May et al. 2020), and/or adjust their height and location so that they don't intrude as much on busy flight paths.

Through a terrestrial utopianism, we can begin to build beyond the mounting socio-ecological perturbations of the Capitalocene and towards what philosopher Glen Albrecht (2020) and others (Prescott & Logan 2017) refer to as the 'Symbiocene'. The concept is rooted in insights from microbiology (see Chapter 2) which underscore the entangled nature of terrestrial life and whose key term literally denotes 'living together'. The Symbiocene would mark a considerable departure from the 'dysbiosis' (Greek etymological roots denoting 'life in distress') that characterises the 'Anthropocene [or Capitalocene] syndrome' (p. 9) by offering a decidedly ecotopian 'vision of a desirable future state toward which social change can be directed' (Albrecht 2020, p. 21). Contrary to Rolston's observation that 'from the skin out ... interactions between individuals are nothing but struggle' (2012, p. 162), which is akin to reducing the real's potential to that which hegemonic systems have designated as 'the way things necessarily are', Albrecht defines this potential new epoch as:

> a period in the history of humanity on this Earth, [which] will be characterized by human intelligence and praxis that replicate the symbiotic and mutually reinforcing life-reproducing forms and processes found in living systems. This period of human existence will be a positive affirmation of life, and it offers the possibility of

the complete re-integration of the human body, psyche and culture with the rest of life. (2019, p. 102)

Albrecht's approach is still rather anthropocentric, affording a little too much primacy to human agency and intellectual prowess; the latter are not, as we've seen, the only or most significant 'Earth' movers. However, Albrecht's emphasis on mutuality and human re-integration with the more-than-human world is an important point, and one that terrestrial ecotopian manifestations emphasise. Moreover, it's important to highlight that terrestrial ecotopianism isn't an abstract theoretical endeavour, nor is it merely alive in the post-anthropocentric sensibilities and modes of relationality evinced by REAs, indigenous communities, and ecotopian literary works discussed in the preceding chapters. Yes, the task at hand is considerable: overhauling how many of us (especially in the West) presently view and relate to other terrestrial kin will require mass collective action for dismantling entrenched anthropocentric paradigms, and an overhaul of growth-oriented socioeconomic systems and their attendant extractive imaginaries that treat terra as a bottomless, inexhaustible pit of exploitable commodities. This in turn will require fundamental, multifarious changes across virtually every aspect of our industrial-capitalist societies – agriculture, transport, housing, energy, and education. In order to accomplish such a Herculean task, we need just about every tool at our disposal, actual and imaginable: mass protests and direct-action, progressive and redistributive government policies, imaginative storytelling, and terrestrial ecotopian experiments in the interstices, at the margins and within as islands of radical possibility amidst a sea of deficient actuality. Often it is such seemingly insignificant experiments, as instantiations of radical love and resistance, that together contribute towards the larger tapestry of transformative change. From rewilding and 'nature' restoration projects cropping up globally (those not reliant on billionaire philanthropic funding) to the spread of 'buzz stops' – barren bus shelter roofs being turned into mini wildlife oases – across Western Europe and North America (Weston 2022). Terrestrial ecotopias are not only alive in mass social mobilisations crying 'no!' to the established order; they can also be seen in such seemingly minute, yet powerful, acts as letting spiders seeking refuge in our home during cold autumn and winter months 'be', glimmers of the possibility of more respectful cohabitation.

To be alive to the world and the multifarious agencies who 'make' it opens up a vast, limitless terrain of joyful possibilities. What a world of difference it makes to see oneself as continually amidst a sea of 'who's' on whom we depend and who depend on us. It alters just about everything – how we interact with, view and attend to others, how we move, nourish ourselves, learn, organise socio-politically, and dream, as well as what we see, hear and feel. It also introduces innumerable extra layers of complexity into our lives, with newfound joys as well as difficulties. Terrestrial ecotopias evoke such a shift in sensibilities and worldviews as one of the indispensable aspects of transformative change for liveable futurities. And what a delight to remember that such transformations are not only possible but are already alive and kicking in the here and now – indeed, they have been for a very long time. The Environmental Justice Atlas website features a 'Blockadia map' that has thus far documented 3,900 cases globally (and counting) of local and indigenous mobilisations for ecological justice and against extractivism. There is no perfection to be had in our relations with others – even love is not without its exclusions and animosities. Love can be messy and painful as it is joyous and wonderful. But love, at least in its positive political modality, 'cherishes otherness' (Blackie 2018, p. 304). It honours the promise of continually striving to do better for the cherished other, to nourish them and help them flourish. Thus, a terrestrial ecotopianism predicated on the ethics and politics of an other-regarding love doesn't merely focus on ensuring the wellbeing of others in light of their indispensability to our own survival; it reminds us to enquire how we can enable our relatives to live well for themselves. It's an imperfect tool, as the best utopian modalities are, but with this as a starting point, alongside collective strivings to stem present tides of ruination and dismantle their systemic foundations, we stand a good chance at evading the worst of all possible futures. Not just this, we might even bring about a renewed earthly vitality previously unimagined. What new life forms and relations might spring forth if colonial-capitalist, anthropocentric, heteronormative and patriarchal violence were substantially diminished or even finally shut down (Gumbs 2021, p. 43)? The future(s) is never written, and thus the fight is far from over.

Bibliography

Aberbach, J. D. & Rockman, B. A. (2002). Conducting and coding elite interviews. *PS: Political Science & Politics*, *35*(4), pp. 673–676.

Adorno, T. W. (1973). *Negative dialectics* (E. B. Ashton, Trans.). Continuum. (Original work published 1966)

Agarwal, A., Narain, S. & Sharma, A. (2017). The global commons and environmental justice – climate change. In *Environmental justice* (pp. 171–199). Routledge.

Agriculture and Horticulture Development Board (AHDB). (2021). *China drives Brazilian beef growth, but for how long?* Access here: https://ahdb.org.uk/news/china-drives-brazilian-beef-export-growth-but-for-how-long

Ajl, M. (2021a). *A people's Green New Deal*. Pluto Press.

Ajl, M. (2021b). A people's Green New Deal: Obstacles and prospects. *Agrarian South: Journal of Political Economy*, *10*(2), pp. 371–390.

Alberro, H. (2019). Methodological considerations for the special-risk researcher: A research note. *Methodological Innovations*, *12*(1), pp. 1–6.

Alberro, H. (2020). Extinction Rebellion: Why disavowing politics is a dead end for climate action. *The Conversation*. Access here: https://theconversation.com/extinction-rebellion-why-disavowing-politics-is-a-dead-end-for-climate-action-145479

Alberro, H. (2021a). In the Shadow of Death: Loss, hope and radical environmental activism in the Anthropocene. *Exchanges: The Interdisciplinary Research Journal*, *8*(2), pp. 8–27.

Alberro, H. (2021b). In and against eco-apocalypse: On the terrestrial ecotopianism of radical environmental activists. *Utopian Studies*, *32*(1), pp. 36–55.

Alberro, H. (2022). HG Wells, earthly and post-terrestrial futures. *Futures*, *140*, pp. 1–9.

Alberro, H. (2024) *Radical environmental resistance: Love, rage and hope in an era of climate and biodiversity breakdown*. Emerald Publishing.

Alberro, H. & Daniele, L. (2021). Ecocide: Why establishing a new international crime would be a step towards interspecies justice. *The Conversation*. Access here: https://theconversation.com/ecocide-why-establishing-a-new-international-crime-would-be-a-step-towards-interspecies-justice-162059

Albrecht, G. A. (2019a). *Earth emotions: New words for a new world*. Ithaca, NY: Cornell University Press.

Albrecht, G. A. (2020). Negating solastalgia: An emotional revolution from the Anthropocene to the Symbiocene. *American Imago*, 77(1), pp. 9–30.

Altvater, E. (2016). The capitalocene, or, geoengineering against capitalism's planetary boundaries. In *Anthropocene or capitalocene*. PM Press, pp. 138–152.

Alves, R. R. N. & Barboza, R. R. D. (2018). What about the unusual soldiers? Animals used in war. In *Ethnozoology* (pp. 323–337). Academic Press.

Andersen, M. S. & Massa, I. (2000). Ecological modernization – origins, dilemmas and future directions. *Journal of Environmental Policy and Planning*, 2(4), pp. 337–345.

Anderson, E. N. (2010). *The pursuit of ecotopia: Lessons from indigenous and traditional societies for the human ecology of our modern world.* ABC-CLIO.

Anderson, M. J. (2009). *Carl Linnaeus: Father of classification.* Enslow Publishing, LLC.

Anderson, K. & Bows, A. (2012). A new paradigm for climate change. *Nature Climate Change*, 2(9), pp. 639–640.

Annette, L. (2022). Establishing a post-2020 global biodiversity framework. *Impact*, 2022(4), pp. 4–5.

Appiah-Opoku, S. (2007). Indigenous beliefs and environmental stewardship: A rural Ghana experience. *Journal of Cultural Geography*, 24(2), pp. 79–98.

Aristotle. (1999). *Politics* (B. Jowett, Trans.). Batoche Books. Retrieved from http://www.efm.bris.ac.uk/het/aristotle/politics.pdf (Original work published 350 BCE).

Armstrong, S. & Botzler, R. (2003). Rene Descartes and animals are machines. In *Environmental ethics: Divergence and convergence* (3rd edn). McGraw New York: McGraw Hill.

Atwood, M. (2014). *The Maddaddam trilogy.* Anchor.

Badiou, A. & Žižek, S. (2009). *Philosophy in the present.* Polity.

Bagchi, B. & Guerra, P. (2016). Many modernities and utopia: From Thomas More to South Asian utopian writings. In P. Guerra (Ed.), *Utopía: 500 años*. Ediciones Universidad Cooperativa de Colombia. http://dx.doi.org/10.16925/9789587600544

Bagchi, B. (2020). Speculating with human rights: Two South Asian women writers and utopian mobilities. *Mobilities*, 15(1), 69–80.

Bahro, R. (1994). *Avoiding social and ecological disaster: The politics of world transformation: An inquiry into the foundations of spiritual and ecological politics.* Gateway Books (GB).

Bak, P. & Sneppen, K. (1993). Punctuated equilibrium and criticality in a simple model of evolution. *Physical review letters*, 71(24), p. 4083.

Baker - Cristales, B. (2012). Poiesis of possibility: The ethnographic sensibilities of Ursula K. Le Guin. *Anthropology and Humanism*, 37(1), pp. 15–26.

Balcombe, J. (2016). *What a fish knows: The inner lives of our underwater cousins.* Scientific American.

Baluška, F. & Mancuso, S. (2018). Plant cognition and behavior: From environmental awareness to synaptic circuits navigating root apices. In *Memory and learning in plants* (pp. 51–77). Springer.

Barnhill, D. L. (2011). Conceiving ecotopia. *Journal for the Study of Religion, Nature & Culture, 5*(2), pp. 145–163.

Bar-On, Y. M., Phillips, R. & Milo, R. (2018). The biomass distribution on Earth. *Proceedings of the National Academy of Sciences, 115*(25), pp. 6506–6511.

Barad, K. (2007). *Meeting the universe halfway: Quantum physics and the entanglement of matter and meaning.* Duke University Press.

Beck, U. (2010). Climate for change, or how to create a green modernity? *Theory, Culture & Society, 27*(2–3), pp. 254–266.

Bellamy, E. (2010). *Looking Backward: 2000–1887.* Penguin Putnam.

Benally, M. D. & Nez Denetdale, J. (2011). *Bitter water: Oral histories of the Navajo-Hopi Land Dispute.* University of Arizona Press.

Bennett, J. (2010). *Vibrant matter: A political ecology of things.* Duke University Press.

Bentham, J. (1948). *An introduction to the principles of morals and legislation.* Hafner. (Original work published 1789)

Berardi, F. B. (2017). *Futurability: The age of impotence and the horizon of possibility.* Verso Books.

Berger, J. (2008). *Hold everything dear: Dispatches on survival and resistance.* Vintage.

Berger, J. (2009). *Why look at animals?* Penguin.

Berny, N. & Rootes, C. (2018). Environmental NGOs at a crossroads? *Environmental Politics, 27*(6), pp. 947–972.

Biro, A. (Ed.). (2011). *Critical ecologies: The Frankfurt School and contemporary environmental crises.* University of Toronto Press.

Blackie, S. (2018). *The enchanted life: Reclaiming the magic and wisdom of the natural world.* September Publishing.

Bloch, E., Plaice, N., Plaice, S. & Knight, P. (1986). *The principle of hope* (Vol. 3, pp. 1938–1947). MIT Press.

Bloch, E. (1995). *The principle of hope* (Vol. 3). The MIT Press.

Bogdanov, A., Graham, L. R. & Stites, R. (1984). *Red Star: The first Bolshevik utopia.* Indiana University Press.

Borrvall, C. & Ebenman, B. (2006). Early onset of secondary extinctions in ecological communities following the loss of top predators. *Ecology Letters, 9*(4), pp. 435–442.

Bötticher, A. (2017). Towards academic consensus definitions of radicalism and extremism. *Perspectives on Terrorism, 11*(4), pp. 73–77.

Boulton, C. A., Lenton, T. M. & Boers, N. (2022). Pronounced loss of Amazon rainforest resilience since the early 2000s. *Nature Climate Change, 12*, pp. 271–278.

Bowen, G. A. (2009). Document analysis as a qualitative research method. *Qualitative research journal*.

Boyd, R. (2022). Making a comeback: Rewilding in Europe gets a £4m funding boost. *The Guardian*. Access here: https://www.theguardian.com/environment/2022/jul/29/making-a-comeback-rewilding-in-europe-gets-a-4m-funding-boost

Breines, W. (1989). *Community and organization in the new left, 1962–1968: The great refusal*. Rutgers University Press.

Brockington, D. (2002). *Fortress conservation: the preservation of the Mkomazi Game Reserve, Tanzania*. Indiana University Press.

Broswimmer, F. J. (2002). *Ecocide: A short history of the mass extinction of species* (p. 105). Pluto Press.

Bruckner, B., Hubacek, K., Shan, Y., Zhong, H. & Feng, K. (2022). Impacts of poverty alleviation on national and global carbon emissions. *Nature Sustainability, 5*, pp. 311–320.

Bachmann, M. E., Junker, J., Mundry, R., Nielsen, M. R., Haase, D., Cohen, H., Kouassi, J. A. & Kuehl, H. S. (2019). Disentangling economic, cultural, and nutritional motives to identify entry points for regulating a wildlife commodity chain. *Biological Conservation, 238*, p. 108177.

Barkun, M. (1986). *Disaster and the millennium*. Syracuse University Press.

Billings, L. (2019). Colonizing other planets is a bad idea. *Futures, 110*, pp. 44–46.

Birch, J., Burn, C., Schnell, A., Browning, H. & Crump, A. (2021). *Review of the evidence of sentience in cephalopod molluscs and decapod crustaceans*. London School of Economics and Political Science. Access here: https://www.lse.ac.uk/News/News-Assets/PDFs/2021/Sentience-in-Cephalopod-Molluscs-and-Decapod-Crustaceans-Final-Report-November-2021.pdf

Bondaroff, P. & Teale, N. (2020). A typology of direct action at sea. In *Non-Human Nature in World Politics* (pp. 279–310). Springer.

Bookchin, M. (1986). *The modern crisis*. New Society Publishers.

Bookchin, M. (2005). *The ecology of freedom*. Oaklands.

Borrows, J. (2002). *Recovering Canada*. University of Toronto Press.

Botz-Bornstein, T. (2012). Critical posthumanism. *Pensamiento y Cultura, 15*(1), pp. 20–30.

Braidotti, R. (2009). Animals, anomalies, and inorganic others. *Pmla, 124*(2), pp. 526–532.

Braidotti, R. (2013). *The posthuman*. Wiley.

Braidotti, R. (2017). Affirmative ethics, sustainable futures. In A. Rodman (Ed.), *Critical and clinical cartographies*. Edinburgh University Press.

Braidotti, R. (2019). A theoretical framework for the critical posthumanities. *Theory, Culture & Society*, *36*(6), pp. 31–61.
Braidotti, R. (2020). 'We' are in this together, but we are not one and the same. *Journal of Bioethical Inquiry*, *17*(4), pp. 465–469.
Chalmers University of Technology. (2022). Agriculture drives more than 90% of tropical deforestation. *ScienceDaily*. Retrieved November 2, 2022, from www.sciencedaily.com/releases/2022/09/220908172312.htm
Clarke, V. & Braun, V. (2013). Successful qualitative research: A practical guide for beginners. In *Successful qualitative research* (pp. 1–400). Sage.
Broom, D. M. (2007). Cognitive ability and sentience: Which aquatic animals should be protected? *Diseases of Aquatic Organisms*, *75*(2), pp. 99–108.
Buber, M. (1996). *Paths in utopia*. Syracuse University Press.
Buber, M. (2012). *I and Thou*. eBookIt.com.
Buhle, P. (2001). Ecotopia. *Capitalism Nature Socialism*, *12*(3), pp. 149–153.
Büscher, B., Fletcher, R., Brockington, D., Sandbrook, C., Adams, W. M., Campbell, L., Corson, C., Dressler, W., Duffy, R., Gray, N. & Holmes, G. (2017). Half-earth or whole earth? Radical ideas for conservation, and their implications. *Oryx*, *51*(3), pp. 407–410.
Büscher, B. & Fletcher, R. (2019). Towards convivial conservation. *Conservation & Society*, *17*(3), 283–296.
Buscher, B. & Fletcher, R. (2020). *The conservation revolution: Radical ideas for saving nature beyond the Anthropocene*. Verso Books.
Butler, J. (2004). *Undoing gender*. Routledge.
Butler, J. (2017). Bodies that matter. In *Feminist theory and the body* (pp. 235–245). Routledge.
Cafaro, P. (2015). Three ways to think about the sixth mass extinction. *Biological Conservation*, *192*, pp. 387–393.
Callenbach, Ernest. (1975). *Ecotopia*. Bantam Books.
Canavan, G. & Robinson, K. S. (Eds). (2014). *Green planets: Ecology and science fiction*. Wesleyan University Press.
Carbon Brief. (2022). Analysis: Do COP26 promises keep global warming below 2C? *Carbon Brief*. Access here: https://www.carbonbrief.org/analysis-do-cop26-promises-keep-global-warming-below-2c
Carmichael, J. T., Brulle, R. J. & Huxster, J. K. (2017). The great divide: Understanding the role of media and other drivers of the partisan divide in public concern over climate change in the USA, 2001–2014. *Climatic Change*, *141*(4), pp. 599–612.
Carrington, D. (2022). Huge UK public support for direct action to protect environment – poll. *The Guardian*. Access here: https://www.theguardian.com/environment/2022/oct/24/huge-uk-public-support-for-direct-action-to-protect-environment-poll

Cassegård, C. & Thörn, H. (2018). Toward a postapocalyptic environmentalism? Responses to loss and visions of the future in climate activism. *Environment and Planning E: Nature and Space*, *1*(4), pp. 561–578.

Cassegård, C. (2023). Activism without hope? Four varieties of postapocalyptic environmentalism. *Environmental Politics*. https://doi.org/10.1080/09644016.2023.2226022

Cathcart, R. S. (1978). Movements: Confrontation as rhetorical form. *Southern Speech Communication Journal*, *43*(3), pp. 233–247.

Ceballos, G. (2016). Four commentaries on the Pope's message on climate change and income inequality. *The Quarterly Review of Biology*, *91*(3), pp. 285–295.

Ceballos, G., Ehrlich, P. R. & Raven, P. H. (2020). Vertebrates on the brink as indicators of biological annihilation and the sixth mass extinction. *Proceedings of the National Academy of Sciences*, *117*(24), pp. 13596–13602.

Celermajer, D., Chatterjee, S., Cochrane, A., Fishel, S., Neimanis, A., O'brien, A., Reid, S., Srinivasan, K., Schlosberg, D. & Waldow, A. (2020). Justice through a multispecies lens. *Contemporary Political Theory*, *19*(3), pp. 475–512.

Chakrabarty, D. (2009). The climate of history: Four theses. *Critical Inquiry*, *35*(2), pp. 197–222.

Chatzidakis, A., Hakim, J., Litter, J. & Rottenberg, C. (2020). *The care manifesto: The politics of interdependence*. Verso Books.

Chesterton, M. (2017). 'The oldest living thing on earth'. BBC News. Access here: https://www.bbc.co.uk/news/science-environment-40224991

Chao, S., Bolender, K. & Kirksey, E. (Eds). (2022). *The promise of multispecies justice*. Duke University Press.

Claeys, G. (2016). *Dystopia: a natural history*. Oxford University Press.

Clark, B. & York, R. (2005). Carbon metabolism: Global capitalism, climate change, and the biospheric rift. *Theory and society*, *34*(4), pp. 391–428.

Climate Action Tracker. (2023a). *Countries*. Available at: https://climateactiontracker.org/countries/

Climate Action Tracker. (2023b). *UAE 2030 climate plan keeps firm focus on fossil fuels*. Available at: https://climateactiontracker.org/press/uae-2030-climate-plan-keeps-firm-focus-on-fossil-fuels/

Cochrane, A. (2009). Ownership and justice for animals. *Utilitas*, *21*(4), pp. 424–442.

Coghlan, S. & Cardilini, A. P. (2022). A critical review of the compassionate conservation debate. *Conservation Biology*, *36*(1), p. e13760.

Collard, R. C., Dempsey, J. & Sundberg, J. (2015). A manifesto for abundant futures. *Annals of the Association of American Geographers*, *105*(2), pp. 322–330.

Convention on Biological Diversity. (2021a). *Kunming declaration.* Access here: https://www.cbd.int/doc/c/df35/4b94/5e86e1ee09bc8c7d4b35aafo/kunmingdeclaration-en.pdf

Convention on Biological Diversity. (2021b). *A new global framework for managing nature through 2030: First detailed draft agreement debuts.* https://www.cbd.int/article/draft-1-global-biodiversity-framework

Corlett, R. T. (2016). Restoration, reintroduction, and rewilding in a changing world. *Trends in Ecology & Evolution, 31*(6), pp. 453–462.

Coulthard, G. (2010) Place against empire: Understanding indigenous anti-colonialism. *Affinities: A Journal of Radical Theory, Culture, and Action, 4*(2), pp. 79–83.

Crane, S. (2021). Francis of Assisi on protecting, obeying, and worshiping with animals. *Exemplaria, 33*(4), pp. 369–388.

Creed, B. (2017). *Stray: Human-animal ethics in the Anthropocene.* Power Publications.

Crenshaw, K. W. (2017). *On intersectionality: Essential writings.* The New Press.

Crist, E. & Kopnina, H. (2014). Unsettling anthropocentrism. *Dialectical Anthropology, 38*(4), pp. 387–396.

Crist, E. (2017). The affliction of human supremacy. *The Ecological Citizen, 1*(1), pp. 61–64.

Cronon, W. (1996). The trouble with wilderness: Or, getting back to the wrong nature. *Environmental History, 1*(1), pp. 7–28.

Crutzen, P. J. (2002). The 'Anthropocene'. *Journal de Physique IV (Proceedings), 12*(10), pp. 1–5.

Crutzen, P. J. (2006). The 'Anthropocene'. In *Earth system science in the Anthropocene* (pp. 13–18). Springer.

Crystal Waters. (2022). *About Crystal Waters.* Access here: https://crystalwaters.org.au/about/

Cudworth, E. & Hobden, S. (2013). Complexity, ecologism, and posthuman politics. *Review of International Studies, 39*(3), pp. 643–664.

Cunsolo, A. & Ellis, N. R. (2018). Ecological grief as a mental health response to climate change-related loss. *Nature Climate Change, 8*(4), pp. 275–281.

da Silva, J. R. (2020). The eco-terrorist wave. *Behavioral Sciences of Terrorism and Political Aggression, 12*(3), pp. 203–216.

Dalton, R. J., Recchia, S. & Rohrschneider, R. (2003). The environmental movement and the modes of political action. *Comparative Political Studies, 36*(7), pp. 743–771.

Darian-Smith, E. (2016). Decolonising utopia. *Australian Journal of Human Rights, 22*(2), pp. 167–183.

Dastur, F. (1996). *Death: An essay on finitude.* A&C Black.

Davis, L. (2012). History, politics, and utopia: Toward a synthesis of social theory and practice. *Existential utopia: New perspectives on utopian thought* (pp. 127–139). Bloomsbury Publishing.

Davis, L. (2021). Grounded utopia. *Utopian Studies, 32*(3), pp. 552–581.

Davis, L. & Kinna, R. (2009). Utopia in contemporary anarchism. Manchester University Press.

Dawson, A. (2016). *Extinction: A radical history*. Or Books.

Day, R. J. (2005). *Gramsci is dead: Anarchist currents in the newest social movements*. Pluto Press.

De la Cadena, M. & Blaser, M. (Eds). (2018). *A world of many worlds*. Duke University Press.

De Cleyre, V. (1912). *Direct action*. Chadwyck-Healey Incorporated.

De Geus, M. (1999). Ecological utopias: Envisioning the sustainable society. *Utopian Studies, 10*(2).

De Waal, F. B. (1999). Anthropomorphism and anthropodenial: Consistency in our thinking about humans and other animals. *Philosophical Topics, 27*(1), pp. 255–280.

De Waal, F. (2016). *Are we smart enough to know how smart animals are?* W. W. Norton.

Deakin, H. & Wakefield, K. (2014). Skype interviewing: Reflections of two PhD researchers. *Qualitative Research, 14*(5), pp. 603–616.

Deleuze, G. & Guattari, F. (2004). *EPZ thousand plateaus*. A&C Black.

Deloria, V. Jr. (2003). *God is red: A native view of religion*. Fulcrum Inc.

Demaria, F. & Kothari, A. (2022). The post-development agenda: Paths to a pluriverse of convivial futures. In F. Adloff & A. Caillé (Eds), *Convivial futures: Views from a post-growth tomorrow* (pp. 139–149). X-Tests on Culture and Society.

Department for Environment, Food and Rural Affairs (DEFRA), Lord Benyon, H. & Lord Goldsmith, H. (2021). *Lobsters, octopuses and crabs recognised as sentient beings*. GOV.UK. Access here: https://www.gov.uk/government/news/lobsters-octopus-and-crabs-recognised-as-sentient-beings

Derrida, J., Magnus, B. & Cullenberg, S. (1994). *Spectres of Marx: The state of the debt, the work of mourning, and the new international*. Routledge.

Derrida, J. (1995a). *The gift of death*. University of Chicago Press.

Derrida, J. (1995b). The time is out of joint. *Deconstruction is/in America: A new sense of the political* (pp. 14–38). NYU Press.

Derrida, J. (1998). Faith and knowledge. In *Religion* (pp. 1–78). Stanford University Press.

Derrida, J. (1999a). *Adieu to Emmanuel Levinas*. Stanford University Press.

Derrida, J. (1999b). Responsabilité et hospitalité. In *Manifeste pour l'hospitalité* (pp. 121–124). Paroles l'Aube.
Derrida, J. (2003a). Autoimmunity: Real and symbolic suicides – a dialogue with Jacques Derrida. In *Philosophy in a time of terror: Dialogues with Jürgen Habermas and Jacques Derrida* (pp. 85–136). University of Chicago Press.
Derrida, J. (2003b). *The work of mourning*. University of Chicago Press
Derrida, J. (2003c). On cosmopolitanism. In *Drifting – architecture and migrancy* (pp. 60–72). Routledge.
Derrida, J. (2005a). Not utopia, the im-possible. In *Paper machine* (pp. 121–135). Stanford University Press.
Derrida, J. (2005b). The principle of hospitality. *parallax*, *11*(1), pp. 6–9.
Derrida, J. (2005c). *Rogues: Two essays on reason*. Stanford University Press.
Derrida, J. (2009). *The beast and the sovereign, Volume I* (Vol. 1; G. Benington, Trans.). Chicago University Press.
Derrida, J. & Wills, D. (2002). The animal that therefore I am (more to follow). *Critical Inquiry*, *28*(2), pp. 369–418.
Despret, V. (2013). From secret agents to interagency. *History and Theory*, *52*(4), pp. 29–44.
Despret, V. (2022) *Living as a bird*. Polity Press.
Descartes, R. (2003). *Discourse on method and related writings* (D. M. Clark, Trans.). Penguin. (Original work published 1637)
Detwiler, F. (1992). 'All my relatives': Persons in Oglala religion. *Religion*, *22*(3), pp. 235–246.
Dick, Philip. K. (2020). Survey team. In M. Ashley (Ed.), *Nature's warnings: Classic stories of eco-science fiction*. The British Library.
Dillon, G. L. (Ed.). (2012). *Walking the clouds: An anthology of Indigenous science fiction*. University of Arizona Press.
Dillon, G. (2016a). Native slipstream. In B. Stratton (Ed.), *The fictions of Stephen Graham: A critical companion*. University of New Mexico Press.
Dinerstein, A. C. (2017). *Concrete utopia: (Re) producing life in, against and beyond the open veins of capital*. Public Seminar. http://www.publicseminar.org/2017/12/concrete-utopia/
Diprose, R. (2006). Derrida and the extraordinary responsibility of inheriting the future-to-come. *Social Semiotics*, *16*(3), pp. 435–447.
Dirzo, R., Young, H. S., Galetti, M., Ceballos, G., Isaac, N. J. & Collen, B. (2014). Defaunation in the Anthropocene. *Science*, *345*(6195), pp. 401–406.
Dobson, A. (2003). *Citizenship and the environment*. OUP Oxford.
Dobson, A. (2010). Democracy and nature: Speaking and listening. *Political Studies*, *58*(4), pp. 752–768.

Domínguez, L. & Luoma, C. (2020). Decolonising conservation policy: How colonial land and conservation ideologies persist and perpetuate indigenous injustices at the expense of the environment. *Land*, *9*(3), pp. 1–22.

Doorly, E. (1942) *The insect man*. Puffin Story Books.

Dow, K. & Lamoreaux, J. (2020). Situated kinmaking and the population 'problem'. *Environmental Humanities*, *12*(2), pp. 475–491.

Dryzek, J. (2007). Political and ecological communication. In J. Dryzek & D. Schlosberg (Eds), *Debating the earth: The environmental politics reader* (pp. 631–646). Oxford University Press.

Duggan, L. & Muñoz, J. E. (2009). Hope and hopelessness: A dialogue. *Women & Performance: A Journal of Feminist Theory*, *19*(2), pp. 275–283.

Dundas, P. (2003). *The Jains*. Routledge.

Dussel, E. (1985). *Philosophy of liberation* (A. Martinez & C. Morkovsky, Trans.). Orbis Books.

Durie, E. T. (2014). Law, responsibility and Maori proprietary interests in water. In *Law, governance and responsibility*. University of Waikato.

Dutton, J. (2010). 'Non-western' utopian traditions. In *The Cambridge companion to utopian literature* (pp. 223–258). Cambridge University Press.

Earth First! UK. (2022). *To our friends from XR: Rebel, Rebel!* Available at: https://www.earthfirst.uk/rebel-rebel/

Earth First! UK Summer Gathering. (2022). Website. Available at: https://www.earthfirst.uk/ef-summer-gathering-2022/

Ehrenreich, B. (2003). Another world is possible. *In These Times*. Access here: https://inthesetimes.com/article/another-world-is-possible

Ehrlich, P. R. & Ehrlich, A. H. (2009). The population bomb revisited. *The Electronic Journal of Sustainable Development*, *1*(3), pp. 63–71.

Eilstrup-Sangiovanni, M. & Bondaroff, T. N. P. (2014). From advocacy to confrontation: Direct enforcement by environmental NGOs. *International Studies Quarterly*, *58*(2), pp. 348–361.

Elden, S. (2006). Heidegger's animals. *Continental Philosophy Review*, *39*(3), pp. 273–291.

Eldredge, N. & Gould, S. J. (1997). On punctuated equilibria. *Science*, *276*(5311), pp. 337–341.

Ellis, E. (2011). The planet of no return: Human resilience on an artificial Earth. *Breakthrough Journal*, *2*(Fall), pp. 37–44.

Ellis, E. C. & Mehrabi, Z. (2019). Half Earth: Promises, pitfalls, and prospects of dedicating Half of Earth's land to conservation. *Current Opinion in Environmental Sustainability*, *38*, pp. 22–30.

Emerman, M. & Malik, H. S. (2010). Paleovirology – modern consequences of ancient viruses. *PLoS Biology*, *8*(2), pp. 1–5.

Engels, F. (1999). *Socialism: Utopian and scientific*. Resistance Books.
Epley, N., Waytz, A. & Cacioppo, J. T. (2007). On seeing human: A three-factor theory of anthropomorphism. *Psychological Review*, *114*(4), p. 864.
Escobar, A. (2020). *Pluriversal politics: The real and the possible*. Duke University Press.
Estes, N. (2019). *Our history is the future: Standing Rock versus the Dakota Access Pipeline, and the long tradition of Indigenous resistance*. Verso.
European Commission. (2022). *Biodiversity strategy for 2030*. Access here: https://environment.ec.europa.eu/strategy/biodiversity-strategy-2030_en#:~:text=The%20EU's%20biodiversity%20strategy%20for,contains%20specific%20actions%20and%20commitments
Federation for Intentional Community. (2022). Access here: https://www.ic.org/
Ferdinand, M. (2021). *Decolonial ecology: Thinking from the Caribbean world*. John Wiley & Sons.
Ferrando, F. (2016). The party of the Anthropocene: Post-humanism, environmentalism and the post-anthropocentric paradigm shift. *Rel.: Beyond Anthropocentrism*, *4*, pp. 159–173.
Firth, R. (2012). *Utopian politics: Citizenship and practice*. Routledge.
Firth, R. (2019). Utopianism and intentional communities. In *The Palgrave handbook of anarchism* (pp. 491–510). Palgrave Macmillan.
Firth, R. & Robinson, A. (2014). For the past yet to come: Utopian conceptions of time and becoming. *Time & Society*, *23*(3), pp. 380–401.
Fisher, M. (2009). *Capitalist realism: Is there no alternative?* John Hunt Publishing.
Food and Agriculture Organisation of the United Nations (FAO). (2022). *The state of the world's forests: Forest pathways for green recovery and building inclusive, resilient and sustainable economies*. Access here: https://www.fao.org/3/cb9363en/cb9363en.pdf
Ford, S. (2003). *Fifty ways to kill a slug*. Octopus.
Forterre, P. (2010). Defining life: The virus viewpoint. *Origins of Life and Evolution of Biospheres*, *40*(2), pp. 151–160.
Foster, J. B., Clark, B. & York, R. (2011). *The ecological rift: Capitalism's war on the earth*. NYU Press.
Foucault, M. & Miskowiec, J. (1986). Of other spaces. *Diacritics*, *16*(1), pp. 22–27.
Foucault, M. (2005). *The order of things*. Routledge.
Frankel, B. (1987). *The post-industrial utopians*. Basil Blackwell.
Fraser, N. (2019). *The old is dying and the new cannot be born: From progressive neoliberalism to Trump and beyond*. Verso Books.
Freud, S. (1984). *On metapsychology: The theory of psychoanalysis*. Penguin Books.
Fromm, E. (2013). *To have or to be?* A&C Black.
Gagliano, M. (2015). In a green frame of mind: Perspectives on the behavioural ecology and cognitive nature of plants. *AoB Plants*, *7*, pp. 1–8.

Garber, D. (1997). Leibniz on form and matter. *Early Science and Medicine*, *2*(3), pp. 326–351.
Garcia-Hernandez, C. C. (2007). Radical environmentalism: The new civil disobedience. *Seattle Journal for Social Justice*, *6*, pp. 289–351.
Garforth, L. (2005). Green utopias: Beyond apocalypse, progress, and pastoral. *Utopian Studies*, *16*(3), pp. 393–427.
Garforth, L. (2009). No intentions? Utopian theory after the future. *Journal for Cultural Research*, *13*(1), pp. 5–27.
Garforth, L. (2018). *Green utopias: Environmental hope before and after nature*. John Wiley & Sons.
Gergen, K. J. (2009). *Relational being: Beyond self and community*. Oxford University Press.
Ghosh, A. (2018). *The great derangement: Climate change and the unthinkable*. Penguin UK.
Ghosh, A. (2021). The Nutmeg's curse. In *The Nutmeg's curse*. University of Chicago Press.
Gilbert, S. F., Sapp, J. & Tauber, A. I. (2012). A symbiotic view of life: We have never been individuals. *The Quarterly Review of Biology*, *87*(4), pp. 325–341.
Glikson, A. (2013). Fire and human evolution: The deep-time blueprints of the Anthropocene. *Anthropocene*, *3*, pp. 89–92.
Global Footprint Network. (2022). *Earth Overshoot Day 2022*. Access here: https://www.overshootday.org/
Goodman, G., Gershwin, M. E. & Bercovich, D. (2014). Mars can wait: Facing the challenges of our civilization. *Israel Medical Association Journal: IMAJ*, *16*(12), pp. 744–747.
Goodpaster, K. (1978). On being morally considerable. *Journal of Philosophy*, *75*, pp. 308–325, p. 310.
Goodwin, B. & Taylor, K. (1982). *The politics of utopia: A study in theory and practice*. St. Martin's Press.
Gorz, A. (1987). *Ecology as politics*. Pluto Press.
Goulson, D. (2021). The insect apocalypse: Our world will grind to a halt without them. *The Guardian*. Access here: https://www.theguardian.com/environment/2021/jul/25/the-insect-apocalypse-our-world-will-grind-to-a-halt-without-them
Goulson, D. (2021b). *Silent earth: Averting the insect apocalypse*. Random House.
Graham, L., Gaulton, R., Gerard, F. & Staley, J. T. (2018). The influence of hedgerow structural condition on wildlife habitat provision in farmed landscapes. *Biological Conservation*, *220*, pp. 122–131.
Gramsci, A. (1971). *Selections from the prison notebooks* (Q. Hoare & G. N. Smith, Ed. & Trans.). International Publishers.

Grasso, D. A., Pandolfi, C., Bazihizina, N., Nocentini, D., Nepi, M. & Mancuso, S. (2015). Extrafloral-nectar-based partner manipulation in plant–ant relationships. *AoB Plants*, *7*, pp. 1–15.

Gray, J. (2007). *Black mass: Apocalyptic religion and the death of utopia*. Macmillan.

Griffin, L. M. (1969). A dramatistic theory of the rhetoric of movements. In *Critical responses to Kenneth Burke* (pp. 456–478). University of Minnesota Press.

Gruen, L. (2015). *Entangled empathy: An alternative ethic for our relationships with animals*. Lantern Books.

Gudynas, E. (2009). La dimensión ecológica del buen vivir: entre el fantasma de la modernidad y el desafío biocéntrico. *OBETS: Revista de Ciencias Sociales*, *4*, pp. 49–54.

Gudynas, E. (2011). Buen Vivir: Today's tomorrow. *Development*, *54*(4), pp. 441–447.

Gumbs, A. P. (2021). *Undrowned: Black feminist lessons from marine mammals*. AK Press.

Karami, A., Golieskardi, A., Keong Choo, C., Larat, V., Galloway, T. S. & Salamatinia, B. (2017). The presence of microplastics in commercial salts from different countries. *Scientific Reports*, *7*(1), pp. 1–11.

Hamilton, C. (2015). Getting the Anthropocene so wrong. *The Anthropocene Review*, *2*(2), pp. 102–107.

Hamilton, C. (2016). The theodicy of the 'Good Anthropocene'. *Environmental Humanities*, *7*(1), pp. 233–238.

Hamilton, L. (2023). Why the ethics of octopus farming are so troubling. *The Conversation*. Available at: https://theconversation.com/why-the-ethics-of-octopus-farming-are-so-troubling-202012#:~:text=Researchers%20have%20suggested%20that%2C%20as,suffering%20on%20an%20unprecedented%20scale.

Haraway, D. (2008). Companion species, mis-recognition, and queer worlding. In N. Giffney & M. J. Hird (Eds), *Queering the non/human* (pp. xxiii–xxxvi). Ashgate.

Haraway, D. J. (2013). *When species meet* (Vol. 3). University of Minnesota Press.

Haraway, D. (2015). Anthropocene, capitalocene, plantationocene, chthulucene: Making kin. *Environmental humanities*, *6*(1), pp. 159–165.

Haraway, D. (2018a). Staying with the trouble for multispecies environmental justice. *Dialogues in Human Geography*, *8*(1), pp. 102–105.

Haraway, D. (2018b). Making kin in the Chthulucene: Reproducing multispecies justice. In *Making kin not population: Reconceiving generations* (pp. 67–99). Prickly Paradigm Press.

Hardt, M. & Negri, A. (2005). *Multitude: War and democracy in the age of empire*. Penguin.

Harman, G. (2009). *Prince of networks: Bruno Latour and metaphysics.* re. press.
Harvey, D. (2007). *A brief history of neoliberalism.* Oxford University Press.
Harvey, D. & Harvey, F. D. (2000). *Spaces of hope* (Vol. 7). University of California Press.
Haynes, J. (1999). Power, politics and environmental movements in the Third World. *Environmental politics, 8*(1), pp. 222–242.
Head, L. (2016). *Hope and grief in the Anthropocene: Re-conceptualising human-nature relations.* Routledge.
Heidegger, M. (1977). *The question concerning technology.* Garland Publishing.
Heidegger, M. (1995). *The fundamental concepts of metaphysics: World, finitude, solitude.* Indiana University Press.
Heise, U. K. (2006). The Hitchhiker's guide to ecocriticism. *Pmla, 121*(2), pp. 503–516.
Heise, U. K. (2013). Globality, difference, and the international turn in ecocriticism. *Pmla, 128*(3), pp. 636–643.
Herrington, R. (2021). Mining our green future. *Nature Reviews Materials, 6*(6), pp. 456–458.
Hester, L. & Cheney, J. (2001). Truth and native American epistemology. *Social Epistemology, 15*(4), pp. 319–334.
Heynen, N. & Van Sant, L. (2015). Political ecologies of activism and direct action politics. In *The Routledge handbook of political ecology* (pp. 169–178). Routledge.
Hickel, J. & Kallis, G. (2020). Is green growth possible? *New Political Economy, 25*(4), pp. 469–486.
Hobbes, T. & Missner, M. (2016). *Thomas Hobbes: Leviathan (Longman library of primary sources in philosophy).* Routledge.
Horton, H. (2023). Fossil fuels received £20bn more UK support than renewables. *The Guardian.* Available at: https://www.theguardian.com/environment/2023/mar/09/fossil-fuels-more-support-uk-than-renewables-since-2015
Hudson, W. D. (1969). Hume on is and ought. In *The is-ought question* (pp. 73–80). Palgrave Macmillan.
Huxley, A. (1963). *The politics of ecology: The question of survival.* Center for the Study of Democratic Institutions.
Huxley, A. (2005) *Island.* Vintage.
Iceland Foods. (2018). *Iceland's banned TV Christmas advert... Say hello to Rang-tan.* YouTube. Access here: https://www.youtube.com/watch?v=JdpspllWI2o
Indigenous Environmental Network (IEN). (2023). *Just transition.* Available at: https://www.ienearth.org/justtransition/
Ingold, T. (2006). Rethinking the animate, re-animating thought. *Ethnos, 71*(1), pp. 9–20.

Intergovernmental Panel on Climate Change (IPCC). (2022). Sixth assessment report: Working group II- Impacts, adaptation and vulnerability. Available at: https://www.ipcc.ch/report/ar6/wg2/downloads/report/IPCC_AR6_WGII_HeadlineStatements.pdf

Intergovernmental Panel on Climate Change (IPCC). (2023). AR6 synthesis report: Climate change 2023. Available at: https://www.ipcc.ch/report/sixth-assessment-report-cycle/

Intergovernmental Science-Policy Platform on Biodiversity and Ecosystem Services. (2019). *Media release: Nature's dangerous decline 'unprecedented'; Species extinction rates 'accelerating'*. Ipbes.net. Access here: https://ipbes.net/news/Media-Release-Global-Assessment#:~:text=The%20Report%20finds%20that%20around,20%25%2C%20mostly%20since%201900.

International Union for the Conservation of Nature (IUCN). (2021). *Deforestation and forest degradation: Issues brief*. Access here: https://www.iucn.org/resources/issues-brief/deforestation-and-forest-degradation

Irving, S. & Helin, J. (2018). A world for sale? An ecofeminist reading of sustainable development discourse. *Gender, Work & Organization*, 25(3), pp. 264–278.

Jakob, M. & Hilaire, J. (2015). Unburnable fossil-fuel reserves. *Nature*, 517(7533), pp. 150–151.

Jameson, F. (1982). Progress versus utopia: Or, can we imagine the future? *Science Fiction Studies*, 9, pp. 147–158.

Jameson, F. (2005). *Archaeologies of the future: The desire called utopia and other science fictions*. Verso.

Jarboe, J. F. (2002). *Testimony*. Federal Bureau of Investigation. Access here: https://archives.fbi.gov/archives/news/testimony/the-threat-of-eco-terrorism

Jazeel, T. (2012). Postcolonialism: Orientalism and the geographical imagination. *Geography*, 97(1), pp. 4–11.

Jeffrey, C. & Dyson, J. (2021). Geographies of the future: Prefigurative politics. *Progress in Human Geography*, 45(4), pp. 641–658.

Jerolmack, C. (2008). How pigeons became rats: The cultural-spatial logic of problem animals. *Social problems*, 55(1), pp. 72–94.

Johnson, E., Morehouse, H., Dalby, S., Lehman, J., Nelson, S., Rowan, R., Wakefield, S. & Yusoff, K. (2014). After the Anthropocene: Politics and geographic inquiry for a new epoch. *Progress in Human Geography*, 38(3), pp. 439–456.

Kanowsky, H. (2022). *Captain Paul Watson cuts ties with Sea Shepherd USA after disagreement of organisation's path*. One Green Planet. Access here: https://www.onegreenplanet.org/animals/captain-paul-watson-cuts-ties-with-sea-shepherd-usa-after-disagreement-of-organizations-path/

Kant, I. & Schneewind, J. B. (2002). *Groundwork for the metaphysics of morals*. Yale University Press.

Kant, I. (2005). Toward perpetual peace. In *Theories of federalism: A reader* (pp. 87–99). Palgrave Macmillan.

Keep it in the Ground. (2022). Over 400 organisations call on world leaders: End fossil fuel development. Access here: http://keepitintheground.org/

Khait, I., Lewin-Epstein, O., Sharon, R., Saban, K., Goldstein, R., Anikster, Y., Zeron, Y., Agassy, C., Nizan, S., Sharabi, G. & Perelman, R. (2023). Sounds emitted by plants under stress are airborne and informative. *Cell, 186*(7), pp. 1328–1336.

Kimmerer, R. (2013a). *Braiding sweetgrass: Indigenous wisdom, scientific knowledge and the teachings of plants.* Milkweed Editions.

Kimmerer, R. W. (2013b). *The democracy of species.* Penguin Books.

Kirkby, J. (2006). 'Remembrance of the future': Derrida on mourning. *Social Semiotics, 16*(3), 461–472.

Kirksey, E., Shapiro, N. & Brodine, M. (2014). Hope in blasted landscapes. In *The multispecies salon* (pp. 25–63). Duke University Press.

Klein, N. (2015). *This changes everything: Capitalism vs. the climate.* Simon and Schuster.

Klein, N. (2020). *On fire: The (burning) case for a green new deal.* Simon & Schuster.

Kleinman, A. (2012). Intra-actions. *Mousse, 34*, pp. 76–81.

Koch, A., Brierley, C., Maslin, M. M. & Lewis, S. L. (2019). Earth system impacts of the European arrival and Great Dying in the Americas after 1492. *Quaternary Science Reviews, 207*, pp. 13–36.

Kolakowski, L. (1983). The death of utopia reconsidered. In *Kolakowski, Modernity on Endless Trial.* University of Chicago.

Kopnina, H. (2012). The Lorax complex: Deep ecology, ecocentrism and exclusion. *Journal of Integrative Environmental Sciences, 9*(4), pp. 235–254.

Korody, M. L., Ford, S. M., Nguyen, T. D., Pivaroff, C. G., Valiente-Alandi, I., Peterson, S. E., Ryder, O. A. & Loring, J. F. (2021). Rewinding extinction in the northern white rhinoceros: Genetically diverse induced pluripotent stem cell bank for genetic rescue. *Stem Cells and Development, 30*(4), pp. 177–189.

Krenak, A. (2023). *Life is not useful.* John Wiley & Sons.

Kröger, M. (2021). *Iron will: Global extractivism and mining resistance in Brazil and India.* University of Michigan Press.

Kropotkin, P. (1908). Syndicalism and anarchism. *Black Flag, 210*, pp. 24–27.

Kuijper, D. P. J., Churski, M., Trouwborst, A., Heurich, M., Smit, C., Kerley, G. I. H. & Cromsigt, J. P. G. M. (2019). Keep the wolf from the door: How to conserve wolves in Europe's human-dominated landscapes? *Biological Conservation, 235*, pp. 102–111.

Kumar, K. (1987). *Utopia and anti-utopia in modern times.* Basil Blackwell.

Kumar, K. (2003). Aspects of the western utopian tradition. *History of the Human Sciences*, *16*(1), pp. 63–77.
Kuper, S. (2015) Thoreau, Leopold, & Carson: Challenging capitalist conceptions of the natural environment. *Consilience*, *13*, pp. 267–283.
Laist, D. W. & Shaw, C. (2006). Preliminary evidence that boat speed restrictions reduce deaths of Florida manatees. *Marine Mammal Science*, *22*(2), p. 472.
Landauer, G. (1978). *For socialism*. Telos Press.
Larsen, S. C. & Johnson, J. T. (2017). *Being together in place: Indigenous coexistence in a more than human world*. University of Minnesota Press.
Latouche, S. (2009). *Farewell to growth*. Polity.
Latour, B. (1993). *The pasteurization of France*. Harvard University Press.
Latour, B. (2004a). *Politics of nature: How to bring the sciences into democracy*. Harvard University Press.
Latour, B. (2004b). Whose cosmos, which cosmopolitics? Comments on the peace terms of Ulrich Beck. *Common knowledge*, *10*(3), pp. 450–462.
Latour, B. (2005). *Reassembling the social: An introduction to actor-network-theory*. Oxford University Press.
Latour, B. (2011). Love your monsters. *Breakthrough Journal*, *2*(11), pp. 21–28.
Latour, B. (2014). Agency at the time of the Anthropocene. *New Literary History*, *45*(1), pp. 1–18.
Latour, B. (2017). *Facing Gaia: Eight lectures on the new climatic regime*. John Wiley & Sons.
Latour, B. (2018). *Down to earth: Politics in the new climatic regime*. John Wiley & Sons.
Latour, B., Milstein, D., Marrero-Guillamón, I. & Rodríguez-Giralt, I. (2018). Down to earth social movements: An interview with Bruno Latour. *Social Movement Studies*, *17*(3), pp. 353–361.
Latour, B. (2021). *After lockdown: A metamorphosis*. Polity Press.
Lazaridis, G., Campani, G. & Benveniste, A. (2016). *The rise of the far right in Europe*. Palgrave Macmillan.
Lazarus, R. S. (1999). Hope: An emotion and a vital coping resource against despair. *Social Research*, *66*, pp. 653–678.
Le Guin, U. (1971). *Earthsea: The first four books*. Penguin Books.
Le Guin, U. (2001). *Tales from Earthsea*. Orion Children's Books.
Le Guin, U. (2012). *The Unreal & the Real*. Gollancz.
Le Guin, U. (2015). *The Word for World is Forest*. Gollancz.
Le Guin, U. (2017). *No Time to Spare*. Houghton Mifflin Publishing Company.
Lehoux, D. (2019). Why does Aristotle think bees are divine? Proportion, triplicity and order in the natural world. *The British Journal for the History of Science*, *52*(3), pp. 383–403.

Leopold, A. (1970). *A Sand County Almanac*. Oxford University Press.
Les Soulèvements de la Terre. Website. Available at: https://lessoulevementsdelaterre.org/
Levinas, E. (1989). *The Levinas reader*. Blackwell.
Levitas, R. (1979). Sociology and utopia. *Sociology, 13*(1), pp. 19–33.
Levitas, R. (1982). Dystopian times? The impact of the death of progress on utopian thinking. *Theory, Culture & Society, 1*(1), pp. 53–64.
Levitas, R. (1990). Educated hope: Ernst Bloch on abstract and concrete utopia. *Utopian Studies, 1*(2), pp. 13–26.
Levitas, R. (2000). For utopia: The (limits of the) utopian function in late capitalist society. *Critical Review of International Social and Political Philosophy, 3*(2–3), pp. 25–43.
Levitas, R. (2003). Introduction: The elusive idea of utopia. *History of the Human Sciences, 16*(1), pp. 1–10.
Levitas, R. A. (2007). The imaginary reconstitution of society: Utopia as method. In *Utopia, method, vision: The use value of social dreaming* (pp. 47–68). Peter Lang International Academic Publishers.
Levitas, R., Sargisson, L., Baccolini, R. & Moylan, T. (2003). *Dark horizons: Science fiction and the dystopian imagination*. Routledge.
Lewis, S., Maslin, M. (2015). Defining the Anthropocene. *Nature*, 519, pp. 171–180. https://doi.org/10.1038/nature14258
Liddick, D. (2006). *Eco-terrorism*. Praeger.
Limbu, S. T. (2020). *Adivasi futurism*. Access here: https://subashthebe.com/wp-content/uploads/2020/08/Adivasi-Futurism-subash-thebe-limbu.pdf
Linkenbach, A. (2009). Doom or salvation? Utopian beliefs in contemporary development discourses and environmentalism. *Sites: A Journal of Social Anthropology and Cultural Studies, 6*(1), pp. 24–47.
Liu, K. (2016). *Invisible planets: 13 visions of the future from China*. Head of Zeus Ltd.
Locke, J. (2000) *Second treatise of government*. Infomotions, Inc. ProQuest Ebook Central, https://ebookcentral.proquest.com/lib/ntuuk/detail.action?docID=3314655
Lockyer, J. & Veteto, J. R. (Eds). (2013). *Environmental anthropology engaging ecotopia: Bioregionalism, permaculture, and ecovillages* (Vol. 17). Berghahn Books.
Lohmann, R. I. (2018). Fiction in fact: Ernest Callenbach's ecotopia and the creation of a green culture with anthropological ingredients. *Anthropology and Humanism, 43*(2), pp. 178–195.
Lorde, A. (2017). *The Master's tools will never dismantle the master's house*. Penguin Random House.
Lorenz, K. (1973). *Civilized man's eight deadly sins*. R. Piper & Co. Verlag.

Lorimer, J., Sandom, C., Jepson, P., Doughty, C., Barua, M. & Kirby, K. J. (2015). Rewilding: Science, practice, and politics. *Annual Review of Environment and Resources*, *40*, pp. 39–62.
Maathai, W. (2010). *The world we once lived in*. Penguin Books.
MacCormack, P. (2020). *The ahuman manifesto: Activism for the end of the Anthropocene*. Bloomsbury Publishing.
Malm, A. & Hornborg, A. (2014). The geology of mankind? A critique of the Anthropocene narrative. *The Anthropocene Review*, *1*(1), pp. 62–69.
Malm, A. (2021). *How to blow up a pipeline*. Verso Books.
Mammola, S., Nanni, V., Pantini, P. & Isaia, M. (2020). Media framing of spiders may exacerbate arachnophobic sentiments. *People and Nature*, *2*(4), pp. 1145–1157.
Mancuso, S. (2021). *The nation of plants: A radical manifesto for humans*. Profile Books.
Mannheim, K. (2013). *Ideology and utopia*. Routledge.
Marcuse, H. (1969). *An essay on liberation* (Vol. 319). Beacon Press.
Marcuse, H. (2013). *One-dimensional man: Studies in the ideology of advanced industrial society*. Routledge.
Marhia, N. (2013). Some humans are more Human than Others: Troubling the 'human' in human security from a critical feminist perspective. *Security Dialogue*, *44*(1), pp. 19–35.
Marx, K. & Engels, F. (2019). The communist manifesto. In *Ideals and ideologies* (pp. 243–255). Routledge.
Mathisen, W. C. (2001). The underestimation of politics in green utopias: The description of politics in Huxley's Island, Le Guin's The Dispossessed, and Callenbach's Ecotopia. *Utopian Studies*, *12*(1), pp. 56–78.
Mauss, M. (1990). *The exchange of gifts* (W. D. Halls, Trans. with Foreword by Mary Douglas). Routledge.
May, R., Nygård, T., Falkdalen, U., Åström, J., Hamre, Ø. & Stokke, B. G. (2020). Paint it black: Efficacy of increased wind turbine rotor blade visibility to reduce avian fatalities. *Ecology and Evolution*, *10*(16), 8927–8935.
Mbembe, A. (2008). Necropolitics. In *Foucault in an Age of Terror* (pp. 152–182). Palgrave Macmillan.
McCarthy, M. (2015). *The Moth Snowstorm: Nature and joy*. New York Review of Books.
McCoyd, J. L. & Kerson, T. S. (2006). Conducting intensive interviews using email: A serendipitous comparative opportunity. *Qualitative Social Work*, *5*(3), pp. 389–406.
McDonald, R. & Beatley, T. (2021). Biophilic cities: Vision and emerging principles. In *Biophilic Cities for an Urban Century* (pp. 63–85). Palgrave Pivot.

McGregor, D., Whitaker, S. & Sritharan, M. (2020). Indigenous environmental justice and sustainability. *Current Opinion in Environmental Sustainability*, *43*, pp. 35–40.
McKibben, B. (2006). *The end of nature*. Random House Trade Paperbacks.
McKinnon, C. (2014). Climate change: Against despair. *Ethics & the Environment*, *19*(1), pp. 31–48.
McNeish, W. (2017). From revelation to revolution: Apocalypticism in green politics. *Environmental Politics*, *26*(6), pp. 1035–1054.
McWethy, D. B., Whitlock, C., Wilmshurst, J. M., McGlone, M. S. & Li, X. (2009). Rapid deforestation of south island, New Zealand, by early Polynesian fires. *The Holocene*, *19*(6), pp. 883–897.
Meadows, D. H., Meadows, D. L., Randers, J. & Behrens III, W. W. (1972). *The limits to growth*. Club of Rome.
Mehrabi, Z. & Naidoo, R. (2022). Shifting baselines and biodiversity success stories. *Nature*, *601*(7894), pp. E17–E18.
Midgley, M. (1998). *Animals and why they matter*. University of Georgia Press.
Midlarsky, M. I. (2011). *Origins of political extremism: Mass violence in the twentieth century and beyond*. Cambridge University Press.
Miéville, C. (2015). The limits of utopia. *Salvage Zone*, *1*.
Milstein, T., McGaurr, L. & Lester, L. (2021). Make love, not war? Radical environmental activism's reconfigurative potential and pitfalls. *Environment and Planning E: Nature and Space*, *4*(2), pp. 296–316.
Minteer, B. A. (2015). The perils of de-extinction. *Minding Nature*, *8*(1), pp. 11–17.
Mohrbacher, B. C. (1996). The whole world is coming: The 1890 ghost dance movement as utopia. *Utopian Studies*, *7*(1), pp. 75–85.
Monash University. (2021). *World's largest study of global climate related mortality links 5 million deaths a year to abnormal temperatures*. Available at: https://www.monash.edu/news/articles/worlds-largest-study-of-global-climate-related-mortality-links-5-million-deaths-a-year-to-abnormal-temperatures
Monbiot, G. (2014). *Feral: Rewilding the land, the sea, and human life*. University of Chicago Press.
Monbiot, G. (2021). After the failure of COP26, there's only one last hope for our survival. *The Guardian*. Access here: https://www.theguardian.com/commentisfree/2021/nov/14/cop26-last-hope-survival-climate-civil-disobedience
Moomaw, W. R., Masino, S. A. & Faison, E. K. (2019). Intact forests in the United States: Proforestation mitigates climate change and serves the greatest good. *Frontiers in Forests and Global Change*, *2*(27), pp. 1–10.
Moore, J. W. (Ed.). (2016). *Anthropocene or capitalocene? Nature, history, and the crisis of capitalism*. PM Press.

Moore, J. W. (2017). The Capitalocene, Part I: On the nature and origins of our ecological crisis. *The Journal of Peasant Studies*, *44*(3), pp. 594–630.
More, T. (1895). *Utopia: Originally printed in Latin, 1516*(14). Constable.
Morin, E. (2007). Restricted complexity, general complexity. In *Science and us: Philosophy and complexity* (pp. 1–25). World Scientific.
Morin, M. E. (2015). The spacing of time and the place of hospitality: Living together according to Bruno Latour and Jacques Derrida. *Parallax*, *21*(1), pp. 26–41.
Morrell, R. & Swart, S. (2005). Men in the third world. In *Handbook of studies on men and masculinities* (pp. 90–113). Sage.
Morton, A. (2022). De-extinction: scientists are planning the multimillion-dollar resurrection of the Tasmanian tiger. *The Guardian*. Access here: https://www.theguardian.com/australia-news/2022/aug/16/de-extinction-scientists-are-planning-the-multimillion-dollar-resurrection-of-the-tasmanian-tiger
Morton, T. (2010). Thinking ecology: The mesh, the strange stranger, and the beautiful soul. *Collapse*, *6*, pp. 265–293.
Morton, T. (2013). *Hyperobjects: Philosophy and ecology after the end of the world*. University of Minnesota Press.
Morton, T. (2017). ... and the Leg Bone's connected to the toxic waste dump bone. *Anthropology of Consciousness*, *28*(2), pp. 135–142.
Moser, K. & Zelaya, K. (Eds). (2020). *The metaphor of the monster: Interdisciplinary approaches to understanding the monstrous other in literature*. Bloomsbury Publishing.
Moylan, T. (1986). *Demand the impossible: Science fiction and the utopian imagination* (Vol. 943). Taylor & Francis.
Moylan, T. & Baccolini, R. (Eds). (2003). *Dark horizons: Science fiction and the dystopian imagination*. Psychology Press.
Moylan, T. (2020). *Becoming utopian: The culture and politics of radical transformation*. Bloomsbury Publishing.
Muldoon, P. (Ed.). (1997). *The Faber book of beasts*. Faber.
Münster, D. (2015). Agrarian alternatives: Agroecology, food sovereignty and the reworking of human-environmental relations in India. *Rivista Degli Studi Orientali* (Rome) Nuova Serie 88 (Supplement), *2*, 233–250.
Murtola, A. M. (2018). *How the global tech elite imagine the future*. Economic and Social Research Aotearoa.
Musk, E. (2018). Elon Musk: We must colonise Mars to preserve our species after a third world war. *The Guardian*. https://www.theguardian.com/technology/video/2018/mar/12/elon-musk-we-must-colonise-mars-to-preserve-our-species-after-a-third-world-war-video#:~:text=Elon%20Musk-,Elon%20Musk%3A%20we%20must%20colonise%20Mars%20to%20preserve%20our%20species,a%20third%20world%20war%20

%E2%80%93%20video&text=Humans%20must% 20prioritise%20the%20colonisation,Elon%20Musk%2C%20said%20on%20Sunday
Mwangi, E. (2019). *The postcolonial animal: African literature and posthuman ethics*. African Perspectives.
Myers, N. (1979). *The sinking Ark. A new look at the problem of disappearing species*. Pergamon Press.
Naess, A. (1995). Self-realization. An ecological approach to being in the world. In G. Sessions (Ed.), *Deep ecology for the twenty-first century* (pp. 225–239). Shambhala.
Naess, A. (2005). The basics of deep ecology. *The Trumpeter, 21*(1).
Nagel, T. (1974). What is it like to be a bat? *Philosophical Review, 83*(4), pp. 435–450.
Nagtzaam, G. & Lentini, P. (2007). Vigilantes on the high seas? The Sea Shepherds and political violence. *Terrorism and Political Violence, 20*(1), pp. 110–133.
Nancy, J. L. (2000). *Being singular plural*. Stanford university press.
Narayanan, Y. & Bindumadhav, S. (2019). 'Posthuman cosmopolitanism' for the Anthropocene in India: Urbanism and human-snake relations in the Kali Yuga. *Geoforum, 106*, pp. 402–410.
NASA Global Climate Change. (2022). *A degree of concern: Why global temperatures matter*. Access here: https://climate.nasa.gov/news/2865/a-degree-of-concern-why-global-temperatures-matter/
Nibert, D. (Ed.). (2017). *Animal oppression and capitalism* (2 Vols). ABC-CLIO.
Nilsen, E. (2019). Read: Alexandra Ocasio-Cortez and Ed Markey's Green New Deal Resolution. *Vox*. Access here: https://www.vox.com/2019/2/7/18215290/alexandria-ocasio-cortez-ed-markey-green-new-deal
Niranjan, A. (2023a). Era of global boiling has arrived, says UN Chief as July set to be hottest month on record. *The Guardian*. Available at: https://www.theguardian.com/science/2023/jul/27/scientists-july-world-hottest-month-record-climate-temperatures
Nuwer, R. (2013). Endangered whales are being sold as dog treats to rich people in Japan. *Smithsonian Magazine*. Access here: https://www.smithsonianmag.com/smart-news/endangered-whales-are-being-sold-as-dog-treats-to-rich-people-in-japan-83670808/
Nyhus, P. J. (2016). Human–wildlife conflict and coexistence. *Annual Review of Environment and Resources, 41*, pp. 143–171.
Ogrodnik, C. & Staggenborg, S. (2016). The ebb and flow of environmentalism. *Sociology Compass, 10*(3), pp. 218–229.
Ojeda, D., Sasser, J. S. & Lunstrum, E. (2020). Malthus's specter and the Anthropocene. *Gender, Place & Culture, 27*(3), pp. 316–332.

Oldekop, J. A., Holmes, G., Harris, W. E. & Evans, K. L. (2016). A global assessment of the social and conservation outcomes of protected areas. *Conservation Biology*, *30*(1), pp. 133–141.
Oppermann, S. & Iovino, S. (2014). Material ecocriticism. In I. Van der Tuin (Ed.), *Gender/nature* (pp. 89–102). Macmillan.
Orrock, J., Connolly, B. & Kitchen, A. (2017). Induced defences in plants reduce herbivory by increasing cannibalism. *Nature Ecology & Evolution*, *1*(8), pp. 1205–1207.
Orwell, G. (1984). *Some thoughts on the common toad*. Penguin Books.
Otto, I. M., Kim, K. M., Dubrovsky, N. & Lucht, W. (2019). Shift the focus from the super-poor to the super-rich. *Nature Climate Change*, *9*(2), pp. 82–84.
Pacoureau, N., Rigby, C. L., Kyne, P. M. et al. (2021). Half a century of global decline in oceanic sharks and rays. *Nature*, 589, pp. 567–571. https://doi.org/10.1038/s41586-020-03173-9
Pacto Ecosocial del Sur. 'Quienes somos?' Pactoecosocialdelsur.com. Access here: https://pactoecosocialdelsur.com/quienes-somos/
Pichler, M., Schaffartzik, A., Haberl, H. & Görg, C. (2017). Drivers of society-nature relations in the Anthropocene and their implications for sustainability transformations. *Current Opinion in Environmental Sustainability*, *26*, pp. 32–36.
Plows, A., Wall, D. & Doherty, B. (2004). Covert repertoires: Ecotage in the UK. *Social Movement Studies*, *3*(2), pp. 199–219.
Pross, A. (2008). How can a chemical system act purposefully? Bridging between life and non‐life. *Journal of Physical Organic Chemistry*, *21*(7–8), pp. 724–730.
Pohl, N. (2019). 'The sense of an ending': Utopia in the Anthropocene. *Књижевна историја – часопис за науку о књижевности*, *51*(167), pp. 65–82.
Popper, K. (2012). *The open society and its enemies*. Routledge.
Popper, K. (2014). *After the open society: Selected social and political writings*. Routledge.
Poore, J. & T. Nemecek. (2018). Reducing food's environmental impacts through producers and consumers. *Science*, *360*(6392), pp. 987–992.
Puniwai, N. (2020). Pua ka wiliwili, nanahu ka manō: Understanding Sharks in Hawaiian Culture. *Human Biology*, *92*(1), pp. 11–17.
Pellow, D. N. (2014). *Total liberation: The power and promise of animal rights and the radical earth movement*. University of Minnesota Press.
Petersen, L. (2010). Reconsidering utopia: On the entanglement of mind and history. *Studies in Social & Political Thought*, *18*, pp. 14–28.
Prescott, S. L. & Logan, A. C. (2017). Down to earth: Planetary health and biophilosophy in the symbiocene epoch. *Challenges*, *8*(2), pp. 1–22.
Pepper, D. (2005). Utopianism and environmentalism. *Environmental Politics*, *14*(1), pp. 3–22.

Pepper, D. (2007). Tensions and dilemmas of ecotopianism. *Environmental Values*, *16*(3), pp. 289–312.
Piccolo, J. J., Taylor, B., Washington, H., Kopnina, H., Gray, J., Alberro, H. & Orlikowska, E. (2022). 'Nature's contributions to people' and peoples' moral obligations to nature. *Biological Conservation*, *270*, p. 109572.
Pike, S. M. (2016). Mourning nature: The work of grief in radical environmentalism. *Journal for the Study of Religion, Nature & Culture*, *10*(4), pp. 419–441.
Piketty, T. & Chancel, L. (2015). Carbon and inequality: From Kyoto to Paris. In *Trends in the global inequality of carbon emissions (1998–2013) and prospects for an equitable adaptation fund*. Paris School of Economics.
Plumptre, A. J., Baisero, D., Belote, R. T., Vázquez-Domínguez, E., Faurby, S., Jędrzejewski, W., Kiara, H., Kühl, H., Benítez-López, A., Luna-Aranguré, C. & Voigt, M. (2021). Where might we find ecologically intact communities? *Frontiers in Forests and Global Change*, *4*, pp. 1–13.
Plumwood, V. (2001). Nature as agency and the prospects for a progressive naturalism. *Capitalism Nature Socialism*, *12*(4), pp. 3–32.
Plumwood, V. (2002). *Feminism and the mastery of nature*. Routledge.
Plumwood, V. (2003). *Animals and ecology: Towards a better integration*. ANU Research Publications.
Plumwood, V. (2007). Journey to the heart of stone. In *Culture, creativity and environment* (pp. 17–36). Brill.
Plumwood, V. (2008). Shadow places and the politics of dwelling. *Australian Humanities Review*, *44*(2), pp. 139–150.
Plumwood, V. (2014). Nature in the active voice. In *The handbook of contemporary animism* (pp. 441–453). Routledge.
Postero, N. (2017). *The indigenous state*. University of California Press.
Pradella, L. (2013). Imperialism and capitalist development in Marx's Capital. *Historical Materialism*, *21*(2), pp. 117–147.
Raupach, M. & Canadell, J. (2010) Carbon and the Anthropocene. *Current Opinion in Environmental Sustainability*, *2*, pp. 210–218.
Regan, T. (2004). *The case for animal rights*. University of California Press.
Reiter, B. (2020). Fuzzy epistemology: Decolonizing the social sciences. *Journal for the Theory of Social Behaviour*, *50*(1), pp. 103–118.
Ricoeur, P. (1986). Ideology and utopia. *Nameh-Ye-Mofid*, *8*(4), pp. 89–104.
Ripple, W. J., Wolf, C., Newsome, T. M., Galetti, M., Alamgir, M., Crist, E., Mahmoud, M. I., Laurance, W. F. & 15,364 Scientist Signatories from 184 Countries. (2017). World scientists' warning to humanity: A second notice. *BioScience*, *67*(12), pp. 1026–1028.
Roberts, M. (2013). Ways of seeing: Whakapapa. *Sites: A Journal of Social Anthropology and Cultural Studies*, *10*(1), pp. 93–120.

Robinson, A. & Tormey, S. (2009). Utopias without transcendence? Post-left anarchy, immediacy and utopian energy. In *Globalization and utopia* (pp. 156–175). Palgrave Macmillan.

Robinson, K. S. (1995). *Pacific Edge*. St Martins Press.

Robinson, K. S. (2020). *Ministry for the future*. Little, Brown Book Group.

Rodman, J. (1980). Paradigm change in political science: An ecological perspective. *American Behavioral Scientist, 24*(1), pp. 49–78.

Rolston, H. & Regan, T. (1988). Life in community: Duties to ecosystems. In *Environmental ethics* (pp. 160–191). Temple University Press. *JSTOR*. http://www.jstor.org/stable/j.ctt1bw1jwf.8. Accessed 7 August 2022.

Rolston, H. (2002). *From beauty to duty: Aesthetics of nature and environmental ethics* (Doctoral dissertation, Colorado State University. Libraries).

Rolston, H. (2015). *After preservation? Dynamic nature in the Anthropocene* (Doctoral dissertation, Colorado State University. Libraries).

Rolston, H. (2020). *A new environmental ethics: The next millennium for life on earth*. Routledge.

Rootes, C. (2014). *Environmental movements: Local, national and global*. Routledge.

Rose, D. B. & Australian Heritage Commission. (1996). *Nourishing terrains: Australian Aboriginal views of landscape and wilderness*. Australian Heritage Commission.

Rose, D. B. (1999). Indigenous ecologies and an ethic of connection. In N. Low (Ed.), *Global ethics and environment* (pp. 175–187). Routledge.

Rose, D. B. (2013). Val Plumwood's philosophical animism: Attentive interactions in the sentient world. *Environmental Humanities, 3*(1), pp. 93–109.

Rose, D. B. (2017). 4. Monk seals at the edge: Blessings in a time of peril. In *Extinction studies* (pp. 117–148). Columbia University Press.

Roy, A. (1999). *The cost of living*. Flamingo.

Rupprecht, C., Cleland, D., Tamura, N., Chaudhuri, R. & Ulibarri, S. (Eds). (2021). *Multispecies cities: Solarpunk urban futures*. World Weaver Press.

Safire, W. (1996). On language: What's an extremist? *The New York Times Magazine*. Access here: https://www.nytimes.com/1996/01/14/magazine/on-language-what-s-an-extremist.html

Said, E. W. (1995). *Orientalism: Western Conceptions of the Orient, with a new Afterword*. Penguin Books.

Salisbury, A. (2022). Pests? Gardeners need to rethink how they view slugs. *The Guardian*. Access here: https://www.theguardian.com/commentisfree/2022/mar/04/pests-gardeners-rethink-slugs-snails-greenfly?CMP=twt_a-environment_b-gdneco&fbclid=IwAR3Zjomw4uDYYJqtJC8F0AqdaUeIWPlhRsEkWNLEHKsnRAL-0GO04XGEr8Q

Samuel, S. (2020). Everywhere basic income has been tried, in one map. *Vox*. Access here: https://www.vox.com/future-perfect/2020/2/19/21112570/universal-basic-income-ubi-map

Sánchez - Bayo, F. & Wyckhuys, K. A. (2021). Further evidence for a global decline of the entomofauna. *Austral Entomology*, *60*(1), pp. 9–26.

Sanders, B. Global Animal Slaughter Statistics and Charts. Faunalytics. (2018). Access here: https://faunalytics.org/global-animal-slaughter-statistics-and-charts/

Sargent, L. T. (1994). The three faces of utopianism revisited. *Utopian Studies*, *5*(1), pp. 1–37.

Sargent, L. T. (2006). In defense of utopia. *Diogenes*, *53*(1), pp. 11–17.

Sargisson, L. (2000). Green utopias of self and other. *Critical Review of International Social and Political Philosophy*, *3*(2–3), pp. 140–156.

Sargisson, L. (2002). *Utopian bodies and the politics of transgression*. Routledge.

Sargisson, L. (2007). Strange places: Estrangement, utopianism, and intentional communities. *Utopian Studies*, *18*(3), pp. 393–424.

Sargisson, L. (2012). *Fool's gold? Utopianism in the twenty-first century*. Springer.

Sargisson, L. (2013). A democracy of all nature: Taking a utopian approach. *Politics*, *33*(2), pp. 124–134.

Sargisson, L. & Sargent, L. T. (2004). *Living in utopia*. Ashgate Publishing.

Sargisson, L. & Sargent, L. T. (2017). *Living in utopia: New Zealand's intentional communities*. Routledge.

Saunders, C. (2012). Reformism and radicalism in the Climate Camp in Britain: Benign coexistence, tensions and prospects for bridging. *Environmental Politics*, *21*(5), pp. 829–846.

Saunders, M. (2022) *This all come back now: An anthology of First Nations speculative fiction*. University of Queensland Press.

Sayers, D. O. (2014). The most wretched of beings in the cage of capitalism. *International Journal of Historical Archaeology*, *18*(3), pp. 529–554.

Schermer, M. H. N. (2007). Brave new world versus island – utopian and dystopian views on psychopharmacology. *Medicine, Health Care and Philosophy*, *10*, pp. 119–128.

Schlosberg, D. & Collins, L. B. (2014). From environmental to climate justice: Climate change and the discourse of environmental justice. *Wiley Interdisciplinary Reviews: Climate Change*, *5*(3), pp. 359–374.

Schultz, P. Wesley, L. Z. & Dalrymple, N. J. (2000). A multinational perspective on the relation between Judeo-Christian religious beliefs and attitudes of environmental concern. *Environment and Behavior*, *32*(4), pp. 576–591.

Schuster, R., Germain, R. R., Bennett, J. R., Reo, N. J. & Arcese, P. (2019) Vertebrate biodiversity on indigenous-managed lands in Australia, Brazil, and Canada equals that in protected areas. *Environmental Science & Policy, 10*, pp. 1–6.

Schwartz, L. (2009). A conversation with Michael Hardt on the politics of love. *Interval (le) s, 3*(1), pp. 810–821.

Scott, C. (2016). (Indigenous) place and time as formal strategy. *Extrapolation, 57*(1–2), pp. 73–94.

Scranton, R. (2015). *Learning to die in the Anthropocene: Reflections on the end of a civilization.* City Lights Publishers.

Serres, M. (1995). *The natural contract.* University of Michigan Press.

Seymour, M. & Wolch, J. (2009). Toward zoöpolis? Innovation and contradiction in a conservation community. *Journal of Urbanism, 2*(3), 215–236.

Shotwell, A. (2016). *Against purity: Living ethically in compromised times.* University of Minnesota Press.

Shue, H. (1993). Subsistence emissions and luxury emissions. *Law & Policy, 15*(1), pp. 39–60.

Shue, H. (2019). Subsistence protection and mitigation ambition: Necessities, economic and climatic. *The British Journal of Politics and International Relations, 21*(2), pp. 251–262.

Simard, S. W. (2018). Mycorrhizal networks facilitate tree communication, learning, and memory. In *Memory and learning in plants* (pp. 191–213). Springer.

Simpson, N. (2021). XR fundamentals: Go beyond politics. *Extinction Rebellion.* Access here: https://rebellion.global/blog/2021/01/05/citizens-assembly-climate-change/

Singer, P. (1973). Animal liberation. In *Animal rights* (pp. 7–18). Palgrave Macmillan.

Singh, V. (2018). What is to be done about climate change? Some thoughts as a writer. "Symposium on the Climate Crisis," *Science Fiction Studies, 45*, pp. 429–430.

Sloam, J., Pickard, S. & Henn, M. (2022). Young people and environmental activism: The transformation of democratic politics. *Journal of Youth Studies, 17*, pp. 1–9.

Smith, N., Bardzell, S. & Bardzell, J. (2017). Designing for cohabitation: Naturecultures, hybrids, and decentering the human in design. In *Proceedings of the 2017 CHI conference on human factors in computing systems* (pp. 1714–1725). Association for Computing Machinery.

Smith, R. K. (2008). Ecoterrorism: A critical analysis of the vilification of radical environmental activists as terrorists. *Environmental Law, 38*, pp. 537–576.

Solnit, R. (2016). *Hope in the dark: Untold histories, wild possibilities.* Haymarket Books.

Sonter, L. J., Ali, S. H. & Watson, J. E. (2018). Mining and biodiversity: Key issues and research needs in conservation science. *Proceedings of the Royal Society B, 285*(1892), pp. 1–9.

Soriano, C. (2018). The Anthropocene and the production and reproduction of capital. *The Anthropocene Review, 5*(2), pp. 202–213.

Sovacool, B. K. (2019). The precarious political economy of cobalt: Balancing prosperity, poverty, and brutality in artisanal and industrial mining in the Democratic Republic of the Congo. *The Extractive Industries and Society, 6*(3), pp. 915–939.

Spiller, C., Erakovic, L., Henare, M. & Pio, E. (2011). Relational well-being and wealth: Māori businesses and an ethic of care. *Journal of Business Ethics, 98*(1), pp. 153–169.

Spivak, G. C. (2005). Scattered speculations on the subaltern and the popular. *Postcolonial Studies, 8*(4), pp. 475–486.

Stamp, E. (2019). Billionaire bunkers: How the 1% are preparing for the apocalypse. *CNN Style*. Access here: https://edition.cnn.com/style/article/doomsday-luxury-bunkers/index.html

Stanescu, J. (2012). Species trouble: Judith Butler, mourning, and the precarious lives of animals. *Hypatia, 27*(3), pp. 567–582.

Starhawk. (1993). *The fifth sacred thing*. Bantam Books.

Steele, W., Wiesel, I. & Maller, C. (2019). More-than-human cities: Where the wild things are. *Geoforum, 106*, pp. 411–415.

Steffen, W., Crutzen, P. J. & McNeill, J. R. (2007). The Anthropocene: Are humans now overwhelming the great forces of nature. *AMBIO: A Journal of the Human Environment, 36*(8), pp. 614–621.

Steffen, W., Persson, Å., Deutsch, L., Zalasiewicz, J., Williams, M., Richardson, K., Crumley, C., Crutzen, P., Folke, C., Gordon, L. & Molina, M. (2011). The Anthropocene: From global change to planetary stewardship. *Ambio, 40*(7), pp. 739–761.

Steffen, W., Broadgate, W., Deutsch, L., Gaffney, O. & Ludwig, C. (2015). The trajectory of the Anthropocene: The great acceleration. *The Anthropocene Review, 2*(1), pp. 81–98.

Steiner, G. (2010). *Anthropocentrism and its discontents: The moral status of animals in the history of western philosophy*. University of Pittsburgh Press.

Stengers, I. (2010). *Cosmopolitics* (Vol. 1). University of Minnesota Press.

Stolze, T. (2018). Against climate stoicism: Learning to fight in the Anthropocene. In *Interrogating the Anthropocene* (pp. 317–337). Palgrave Macmillan.

Strauss, W. (2010). *The myth of endless growth: Exposing capitalism's insustainability*. Lulu.com.

Stuart, A., Thomas, E. F., Donaghue, N. & Russell, A. (2013). 'We may be pirates, but we are not protesters': Identity in the Sea Shepherd Conservation Society. *Political Psychology*, 34(5), pp. 753–777.
Swift, J., Davis, H. & Williams, H. (1959). *Gulliver's travels: 1726* (p. 259). Blackwell.
Suchet - Pearson, S., Wright, S., Lloyd, K., Burarrwanga, L. & Bawaka Country. (2013). Caring as country: Towards an ontology of co - becoming in natural resource management. *Asia Pacific Viewpoint*, 54(2), pp. 185–197.
Summer, S. (2022). *Endless forms: The secret world of wasps*. William Collins.
Sunrise Movement. (2021). *What we believe: Sunrise's principles*. Access here: www.sunrisemovement.org/principles/
Suvin, D. (1988). Not only but also: On cognition and ideology in SF and SF criticism. In *Positions and presuppositions in science fiction* (pp. 44–60). Palgrave Macmillan.
Suvin, D. (1990). Locus, horizon, and orientation: The concept of possible worlds as a key to utopian studies. *Utopian Studies*, 1(2), pp. 69–83.
Suvin, D. (2000). Novum is as novum does. In K. Sayer & J. Moore (Eds), *Science fiction, critical frontiers* (pp. 3–22). Palgrave Macmillan.
Suvin, D. (2021). Utopia or bust: Capitalocene, method, anti-utopia. *Utopian Studies*, 32(1), pp. 1–35.
Tarrow, S. (2013). *The language of contention: Revolutions in words, 1688–2012*. Cambridge University Press.
Taylor, N. (2012). Animals, mess, method: Post-humanism, sociology and animal studies. In L. Birke & J. Hockenhull (Eds), *Crossing boundaries: Investigating human-animal relationships* (pp. 37–50). BRILL.
Taylor, K. (2013). *Political ideas of the utopian socialists*. Routledge.
Taylor, Z. J. & Aalbers, M. B. (2022). Climate gentrification: Risk, rent, and restructuring in Greater Miami. *Annals of the American Association of Geographers*, 112(6), pp. 1685–1701.
Te Aho, L. (2019). Te Mana o te Wai: An indigenous perspective on rivers and river management. *River Research and Applications*, 35(10), pp. 1615–1621.
Temper, L. (2019). Radical climate politics: From Ogoniland to Ende Gelände. In *Routledge handbook of radical politics* (pp. 97–106). Routledge.
Temperton, J. (2017). Now I am become death, the destroyer of worlds. The story of Oppenheimer's infamous quote. *Wired*. Access here: https://www.wired.co.uk/article/manhattan-project-robert-oppenheimer
The Guardian. (2022). Big cats, big cities: How Los Angeles and Mumbai live cheek by jowl with feline locals. Available here: https://www.theguardian.com/environment/2022/jun/30/los-angeles-mumbai-mountain-lion-leopard?CMP=twt_a-environment_b-gdneco&fbclid=IwAR3gxfewLiKqI1zz6I FVtmRTzIBjTYUaASodEsrFH_PT5wjTMFGCo_Kd66hM

Thematic Social forum (TSF). (2012). *Another future is possible*. Access here: http://rio20.net/wp-content/uploads/2012/02/Another-Future-is-Possible_english_web.pdf

Thomas, L. (2022) *The intersectional environmentalist: How to dismantle systems of oppression to protect people and planet*. Souvenir Press.

Thompson, P. & Žižek, S. (Eds). (2014). *The privatization of hope: Ernst Bloch and the future of utopia, SIC 8* (Vol. 8). Duke University Press.

Tilly, C. & Tarrow, S. G. (2015). *Contentious politics*. Oxford University Press.

Timberg, S. (2008). The novel that predicted Portland. *The New York Times*. Access here: https://www.nytimes.com/2008/12/14/fashion/14ecotopia.html?

Toncheva, S. & Fletcher, R. (2021). From conflict to conviviality? Transforming human-bear relations in Bulgaria. *Frontiers in Conservation Science*, *2*, pp. 1–15.

Treanor, B. (2019). Hope in the age of the Anthropocene. *Analecta Hermeneutica*, *10*, pp. 1–22.

Tschakert, P., Schlosberg, D., Celermajer, D., Rickards, L., Winter, C., Thaler, M., Stewart - Harawira, M. & Verlie, B. (2021). Multispecies justice: Climate - just futures with, for and beyond humans. *Wiley Interdisciplinary Reviews: Climate Change*, *12*(2), pp. 1–10.

Turnbaugh, P. J., Ley, R. E., Hamady, M., Fraser-Liggett, C. M., Knight, R. & Gordon, J. I. (2007). The human microbiome project. *Nature*, *449*(7164), pp. 804–810.

UK Student Climate Network. (2022). *The Green New Deal*. Access here: https://ukscn.org/the-green-new-deal/

Ulmer, J. B. (2017). Posthumanism as research methodology: Inquiry in the Anthropocene. *International Journal of Qualitative Studies in Education*, *30*(9), pp. 832–848.

UNEP. (2021). *Making peace with nature: A scientific blueprint to tackle the climate, biodiversity and pollution emergencies*. Access here: https://wedocs.unep.org/xmlui/bitstream/handle/20.500.11822/34949/MPN_ESEN.pdf

UNICEF. (2023). *Devastating floods in Pakistan*. Available at: https://www.unicef.org/emergencies/devastating-floods-pakistan-2022

Union of Concerned Scientists. (1992). *1992 world scientists' letter to humanity*. Ucsusa.org. Access here: https://www.ucsusa.org/resources/1992-world-scientists-warning-humanity

Van der Heijden, H. A. (2006). Globalization, environmental movements, and international political opportunity structures. *Organization & Environment*, *19*(1), pp. 28–45.

Van Dooren, T. (2014a). Mourning crows: Grief and extinction in a shared world. In *Routledge handbook of human-animal studies* (pp. 293–307). Routledge.

Van Dooren, T. (2014b). Care. *Environmental Humanities*, *5*(1), pp. 291–294.

Van Klink, P. (2021). *The utopia of Rojava: A new world for stateless people*. This is a pre-copyedited, author-produced version of an article accepted for publication in Oñati Socio-legal Series. http://opo.iisj.net/index.php/osls/index; following peer review.

Verne, J. (2017). *Twenty thousand leagues under the sea*. Penguin Classics.

Vimal, R., Navarro, L. M., Jones, Y., Wolf, F., Le Moguédec, G. & Réjou-Méchain, M. (2021). The global distribution of protected areas management strategies and their complementarity for biodiversity conservation. *Biological Conservation*, *256*, pp. 1–9.

Vizenor, G. (Ed.). (2008). *Survivance: Narratives of native presence*. University of Nebraska Press.

Wake, D. B. & Vredenburg, V. T. (2008). Are we in the midst of the sixth mass extinction? A view from the world of amphibians. *Proceedings of the National Academy of Sciences*, *105*(Supplement 1), pp. 11466–11473.

Wallach, A. D., Batavia, C., Bekoff, M., Alexander, S., Baker, L., Ben‑Ami, D., Boronyak, L., Cardilli, A. P., Carmel, Y., Celermajer, D. & Coghlan, S. (2020). Recognizing animal personhood in compassionate conservation. *Conservation Biology*, *34*(5), pp. 1097–1106.

Ware, O. (2004). Dialectic of the past/disjuncture of the future: Derrida and Benjamin on the concept of Messianism. *Journal for Cultural and Religious Theory*, *5*(2), pp. 99–114.

Washington, W., Taylor, B., Kopnina, H., Cryer, P. & Piccolo, J. J. (2017). Why ecocentrism is the key pathway to sustainability. *Ecological Citizen*, *1*(1), pp. 35–41.

Washington, H., Piccolo, J., Gomez-Baggethun, E., Kopnina, H. & Alberro, H. (2021). The trouble with anthropocentric hubris, with examples from conservation. *Conservation*, *1*(4), pp. 285–298.

Waters, C. N., Zalasiewicz, J., Summerhayes, C., Barnosky, A. D., Poirier, C., Gałuszka, A., Cearreta, A., Edgeworth, M., Ellis, E. C., Ellis, M. & Jeandel, C. (2016). The Anthropocene is functionally and stratigraphically distinct from the Holocene. *Science*, *351*(6269), p. 137.

Watson, A. & Huntington, O. H. (2008). They're here – I can feel them: the epistemic spaces of Indigenous and Western knowledges. *Social & Cultural Geography*, *9*(3), pp. 257–281.

Webb, D. (2016). Educational studies and the domestication of utopia. *British Journal of Educational Studies*, *64*(4), pp. 431–448.

Weitzenfeld, A. & Joy, M. (2014). An overview of anthropocentrism, humanism, and speciesism in critical animal theory. *Counterpoints*, *448*, pp. 3–27.

Wells, H. G. (1913). *The discovery of the future* (No. 25). BW Huebsch.

Wells, H. G. (1923). *Men like gods*. W. Collins Sons & Co.

Wells, H. G. (1945). *Mind at the end of its tether*. Windmill Press.

Wells, H. G. (2005). *A modern utopia*. Penguin Classics. (Original work published 1905)

Westmoreland, M. W. (2008). Interruptions: Derrida and hospitality. *Kritike*, 2(1), pp. 1–10.

Weston, P. (2022). Buzz stops: Bus shelter roofs turned into gardens for bees and butterflies. *The Guardian*. Access here: https://www.theguardian.com/environment/2022/sep/24/bus-shelter-roofs-turned-into-gardens-for-bees-butterflies-aoe

White, L. (1967). The historical roots of our ecologic crisis. *Science*, 155(3767), pp. 1203–1207.

Whitehead, J. (Ed.). (2020). *Love after the end: An anthology of two-spirit and indigiqueer speculative fiction*. Arsenal Pulp Press.

Whyte, D. (2023). *Carbon cash machine*. Center for Climate Crime and Climate Justice. Available at: https://ccccjustice.org/2023/08/08/carbon-cash-machine/

Whyte, K. (2018). Critical investigations of resilience: A brief introduction to indigenous environmental studies & sciences. *Daedalus*, 147(2), pp. 136–147.

Williamson, M. (2016). Can Musk achieve his Mars dream? *Engineering & Technology*, 11(10), pp. 18–19.

Willow, A. J. (2016). Indigenous ExtrACTIVISM in Boreal Canada: Colonial legacies, contemporary struggles and sovereign futures. *Humanities*, 5(3), pp. 1–15.

Willow, A. J. & Wylie, S. (2014). Politics, ecology, and the new anthropology of energy: Exploring the emerging frontiers of hydraulic fracking. *Journal of Political Ecology*, 21(1), pp. 222–236.

Wilson, E. O. (1984). Biophilia. In *Biophilia*. Harvard University Press.

Wilson, E. O. (2016). *Half-earth: Our planet's fight for life*. W. W. Norton & Company.

Win, T. L. (2020). G20 countries still backing fossil fuels through COVID-19 response. *Reuters*. Access here: https://www.reuters.com/article/us-g20-climatechange-energy-trfn-idUSKBN27Q00Q

Wisman, J. D. (2011). Inequality, social respectability, political power, and environmental devastation. *Journal of Economic Issues*, 45(4), pp. 877–900.

Wolfe, C. (1998). *Old orders for new: Ecology, animal rights, and the poverty of humanism*. Johns Hopkins University Press.

Wolfe, C. (2017). *Extinction studies: Stories of time, death, and generations*. Columbia University Press.

Wolloch, N. (2017). *Subjugated animals: Animals and anthropocentrism in early modern European culture*. Prometheus Books.

Walker, P. & Allegretti, A. (2022). Jacob Rees-Mogg dismisses 'hysteria' over fracking as ban ends. *The Guardian*. Access here: https://www.theguardian.com/envi

ronment/2022/sep/22/government-confirms-it-is-lifting-ban-on-fracking-in-england

World Business Council for Sustainable Development (WBCSD). (2021). *Vision 2050: Time to transform*. Access here: https://timetotransform.biz/wp-content/uploads/2021/03/WBCSD_Vision_2050_Time-To-Transform.pdf

World Meteorological Association (WMO). (2022). 'More bad news for the planet: Greenhouse gas levels hit new highs.' Access here: https://public.wmo.int/en/media/press-release/more-bad-news-planet-greenhouse-gas-levels-hit-new-highs

World Wildlife Fund (WWF). (2022). *Living Planet Report 2022*. Access here: https://livingplanet.panda.org/en-GB/

Wretched of the Earth. (2019). *An open letter to Extinction Rebellion*. Available at: https://www.redpepper.org.uk/an-open-letter-to-extinction-rebellion/

Wright, T. (2012). *The political economy of the Chinese coal industry: Black gold and blood-stained coal*. Routledge.

Yazzie, M. K. (2018). Decolonizing development in Diné Bikeyah: Resource extraction, anti-capitalism, and relational futures. *Environment and Society*, 9(1), pp. 25–39.

Yin, J. (2018). Beyond postmodernism: A non-western perspective on identity. *Journal of Multicultural Discourses*, 13(3), pp. 193–219.

Yin, J., Zhu, S., MacNaughton, P., Allen, J. G. & Spengler, J. D. (2018). Physiological and cognitive performance of exposure to biophilic indoor environment. *Building and Environment*, 132, pp. 255–262.

Young, H. S., McCauley, D. J., Galetti, M. & Dirzo, R. (2016). Patterns, causes, and consequences of Anthropocene defaunation. *Annual Review of Ecology, Evolution, and Systematics*, 47(1), pp. 333–358.

Youngblood, M. (2020). Extremist ideology as a complex contagion: The spread of far-right radicalization in the United States between 2005 and 2017. *Humanities and Social Sciences Communications*, 7(1), pp. 1–10.

Yusoff, K. (2018). *A billion black Anthropocenes or none*. University of Minnesota Press.

Ziemke, T. (2016). The body of knowledge: On the role of the living body in grounding embodied cognition. *Biosystems*, 148, pp. 4–11.

Zalasiewicz, J., Waters, C. N. & Williams, M. (2014). Human bioturbation, and the subterranean landscape of the Anthropocene. *Anthropocene*, 6, pp. 3–9.

Zalasiewicz, J., Waters, C. N., Williams, M., Barnosky, A. D., Cearreta, A., Crutzen, P., Ellis, E., Ellis, M. A., Fairchild, I. J., Grinevald, J. & Haff, P. K. (2015). When did the Anthropocene begin? A mid-twentieth century boundary level is stratigraphically optimal. *Quaternary International*, 383, pp. 196–203.

Zamyatin, Y. (1993) *We*. Penguin Books.

Zerbini, A. N., Adams, G., Best, J., Clapham, P. J., Jackson, J. A. & Punt, A. E. (2019). Assessing the recovery of an Antarctic predator from historical exploitation. *Royal Society Open Science*, *6*(10), pp. 1–22.

Žižek, S. (2011). *Living in the end times*. Verso Books.

Zografos, C. & Robbins, P. (2020). Green sacrifice zones, or why a green new deal cannot ignore the cost shifts of just transitions. *One Earth*, *3*(5), pp. 543–546.

Index

Always Coming Home (Le Guin) 121–123, 126, 129–130, 141, 198, 201, 204, 227
   see also Le Guin
Anthropocene 11, 22–25, 30–32, 34
Anthropocentrism / Western Anthropocentrism 35, 37–39, 41, 45, 47–49, 51, 65–66, 98–99, 138, 147, 164, 171, 187

Berger, John 19, 41, 45–46, 49, 84, 105, 138, 200, 202, 215, 248
biocentrism 57–59
Bloch, Ernst 71–73, 75, 82, 86, 91, 178, 222
Braidotti, Rosi 50, 51, 53, 61, 62, 79, 97–98, 133, 153, 158–159, 185
Buen Vivir 238

Callenbach, Ernest 93, 110, 111, 115–117, 127
Capitalocene, The 25–27, 30, 33–34, 49, 61, 84, 91, 101, 161, 166
Chthulucene 35, 225
   see also Haraway
Climate Action Tracker 215–216
Clive, Hamilton 32
colonialism 27, 38, 41, 65, 104, 185, 201, 204
convivial conservation 232–233, 236
Critical posthumanism 97–99, 102–103
   see also Braidotti
'critical modality of hope' (Duggan and Muñoz) 165–166, 178, 222
'critical utopia' 87–88

   see also Moylan

degrowth 241
Derrida, Jacques 38, 41, 46, 65–66, 78, 79–81, 99, 166, 179–180, 219–22, 224–226, 230, 247
Descartes, René 40
Dillon, Grace 202, 204
Diné 199
Direct Action (DA) 148–149

Earth First! (EF!) 152–153, 161, 164, 175
ecocentrism 57–59
ecocriticism 109
ecological animalism 123–125
   see also Le Guin
Ecotopia (Callenbach) 115–118
   see also Callenbach
ecotopia(s), green utopianism 92–95
Environmental Justice Atlas 251
Erle, Ellis 31–32
Extinction Rebellion 175, 180

Fabre, Jean-Henri 48, 61
'fortress conservation' 169

Garforth, Lisa 245
Ghosh, Amitav 29, 33, 110, 198
Ghost Dance 204
(The) Great Acceleration 24, 28
Green New Deal 143, 238–240, 242–243
'green sacrifice zones' 240

'half-earth' 172

Haraway, Donna 24, 29, 34, 35, 41, 64, 88, 102–103, 189
heterotopia 89–90
Honourable Harvest 193–195, 196
   see also Kimmerer
Huxley, Aldous 93, 95, 110, 111–114, 163

Indigenous: peoples, worldviews, ethics, cosmologies, ecotopias 183–187
Industrial capitalism 27, 28, 32, 38, 49–50, 87, 97, 138, 158, 187
   colonial-capitalism 26, 33, 150, 171, 183
   Western-capitalism 29, 160, 204, 238
Island (Huxley) 111-14, 160, 228, 247
   see also Huxley

Just Stop Oil (JSO) 153–154

Kant, Immanuel 42, 136
Kimmerer, Robin Wall 190, 194–195, 197

Latour, Bruno 55, 62–63–64, 76, 79, 84, 86, 99–101, 104–105, 130, 136–138, 155–156, 161–162, 167, 173, 177, 179–180, 198, 223–225, 229–231
'limits-to-growth' 29, 93, 112, 158
literary ecotopias 109–111
Le Guin, Ursula 50, 52–53, 56, 89, 93, 110, 111, 120–123, 127, 140, 162, 194, 197, 218, 231, 233
Les Soulèvements de la Terre 157
Love After the End (Whitehead) 183, 205–209, 212–213

Malm, Andreas 30, 147, 149, 215–216
Malthusian 29, 116
Māori 192–193
Marx, Karl 50, 76–77, 84

Morton, Timothy 17, 63–64, 79–80, 162, 168, 173
Moylan, Tom 40, 70, 87–88, 103, 106, 109, 118–119, 121, 140, 157, 177
multispecies justice 102–103
Multispecies cities: Solarpunk Urban Futures (Rupprecht et al) 110, 111, 131–140

Pacific Edge (Robinson) 118–120
   see also Robinson
Paris Agreement 12, 13, 16, 17
Plumwood, Val 40, 44–45, 54–55, 98–99, 101, 123–124, 155–156, 167, 173, 176, 187, 194, 223–224, 231
post-anthropocentrism 153, 155–156, 178

'radical flank effect' 147
radical environmental activists (REAs) 96, 98, 144–150, 186, 222, 225
Robinson, Kim Stanley 93, 110, 118–120, 127, 140, 218

Sea Shepherd Conservation Society (SSCS) 154
sixth mass extinction 20, 21
   see also WWF Living Planet Report
Solnit, Rebecca 2, 75, 88, 166, 178
Starhawk 71, 93, 106, 110, 111, 140, 209
Sustainable Development Goals 171
Symbiocene 249–250

terrestrial ecotopianism 157, 177, 179, 186, 223, 245–251
terrestrial kin 41, 167, 233, 242
The Fifth Sacred Thing (Starhawk) 126–131, 141, 195, 198, 201, 204, 209
   see also Starhawk
This All Come Back Now (Saunders) 184, 209–211, 212–213

utopia, utopianism 69–78, 106, 178

Van Dooren, Thom 63–64, 135–136, 156, 219–222

Wells, H.G 1, 95–96, 201

Western Humanism 38
White, Lynn 39
WWF Living Planet Report 20–22

Yolŋu 191

Ralahine Utopian Studies

Ralahine Utopian Studies is the publishing project of the Ralahine Centre for Utopian Studies at the University of Limerick in association with the University of Bologna, the University of Cyprus, and the University of Florida.

The series publishes high-quality scholarship that addresses the theory and practice of utopianism (including Anglophone, continental European and indigenous and postcolonial traditions, and contemporary and historical periods). Publications (in English and other European languages) include original monographs and essay collections (including theoretical, textual and ethnographic/institutional research), English-language translations of utopian scholarship in other national languages, reissues of classic scholarly works that are out of print and annotated editions of original utopian literary and other texts (including translations).

While the series editors seek work that engages with the current scholarship and debates in the field of utopian studies, they will not privilege any particular critical or theoretical orientation. They welcome submissions by established or emerging scholars working within or outside the academy. Given the multilingual and interdisciplinary remit of the series, the editors especially welcome comparative studies in any disciplinary or transdisciplinary framework.

Those interested in contributing to the series are invited to submit a detailed project outline to one of the series editors listed below.

Email queries can also be sent to ireland@peterlang.com.

Series editors:
Raffaella Baccolini (University of Bologna)
Antonis Balasopoulos (University of Cyprus)
Joachim Fischer (University of Limerick)
Michael G. Kelly (University of Limerick)
Tom Moylan (University of Limerick)
Phillip E. Wegner (University of Florida)

Ralahine Centre for Utopian Studies, University of Limerick
http://www3.ul.ie/ralahinecentre/

Volume 1   Tom Moylan and Raffaella Baccolini (eds):
            Utopia Method Vision. The Use Value of Social Dreaming.
            343 pages. 2007. ISBN 978-3-03910-912-8

Volume 2	Michael J. Griffin and Tom Moylan (eds): Exploring the Utopian Impulse. Essays on Utopian Thought and Practice. 434 pages. 2007. 408 pages. 2015. ISBN 978-3-03910-913-5
Volume 3	Ruth Levitas: The Concept of Utopia. (Ralahine Classic) 280 pages. 2010. ISBN 978-3-03911-366-8
Volume 4	Vincent Geoghegan: Utopianism and Marxism. (Ralahine Classic) 189 pages. 2008. ISBN 978-3-03910-137-5
Volume 5	Barbara Goodwin and Keith Taylor: The Politics of Utopia. A Study in Theory and Practice. (Ralahine Classic) 341 pages. 2009. ISBN 978-3-03911-080-3
Volume 6	Darko Suvin: Defined by a Hollow. Essays on Utopia, Science Fiction and Political Epistemology. (Ralahine Reader) 616 pages. 2010. ISBN 978-3-03911-403-0
Volume 7	Andrew Milner (ed.): Tenses of Imagination. Raymond Williams on Science Fiction, Utopia and Dystopia. (Ralahine Reader) 253 pages. 2010. ISBN 978-3-03911-826-7
Volume 8	Nathaniel Coleman (ed.): Imagining and Making the World. Reconsidering Architecture and Utopia. 393 pages. 2011. ISBN 978-3-0343-0120-6
Volume 9	Henry Near: Where Community Happens. The Kibbutz and the Philosophy of Communalism. 256 pages. 2011. ISBN 978-3-0343-0133-6
Volume 10	Robert C. Elliott: The Shape of Utopia. Studies in a Literary Genre. Edited with an Introduction by Phillip E. Wegner. (Ralahine Classic) 170 pages. 2013. ISBN 978-3-0343-0772-7
Volume 11	Michael E. Gardiner: Weak Messianism. Essays in Everyday Utopianism. 284 pages. 2013. ISBN 978-3-0343-0716-1
Volume 12	Matthew Beaumont: The Spectre of Utopia. Utopian and Science Fictions at the Fin de Siècle. 319 pages. 2012. ISBN 978-3-0343-0725-3

Volume 13   Artur Blaim:
Gazing in Useless Wonder. English Utopian Fictions, 1516 –1800.
366 pages. 2013. ISBN 978-3-0343-0899-1

Volume 14   Tom Moylan:
Demand the Impossible. Science Fiction and the Utopian Imagination.
Edited by Raffaella Baccolini. (Ralahine Classic)
358 pages. 2014. ISBN 978-3-0343-0752-9

Volume 15   Phillip E. Wegner:
Shockwaves of Possibility. Essays on Science Fiction, Globalization, and Utopia.
328 pages. 2014. ISBN 978-3-0343-0741-3

Volume 16   Angelika Bammer:
Partial Visions. Feminism and Utopianism in the 1970s.
(Ralahine Classic)
408 pages. 2015. ISBN 978-3-0343-0897-7

Volume 17   Edward K. Chan:
The Racial Horizon of Utopia. Unthinking the Future of Race in Late Twentieth-Century American Utopian Novels.
226 pages. 2016. ISBN 978-3-0343-1916-4

Volume 18   Darko Suvin:
Metamorphoses of Science Fiction. On the Poetics and History of a Literary Genre.
Edited by Gerry Canavan. (Ralahine Classic)
515 pages. 2016. ISBN 978-3-0343-1948-5

Volume 19   Michael S. Cummings:
Children's Voices in Politics.
554 pages. 2020. ISBN 978-3-0343-1943-0

Volume 20   Maïté Maskens and Ruy Blanes (eds):
Utopian Encounters. Anthropologies of Empirical Utopias.
246 pages. 2018. ISBN 978-1-78707-247-3

Volume 21   Peter Fitting:
Utopian Effects, Dystopian Pleasures.
Edited by Brian Greenspan.
458 pages. 2020. ISBN 978-1-78874-353-2

Volume 22   Raffaella Baccolini and Lyman Tower Sargent (eds):
Transgressive Utopianism: Essays in Honor of Lucy Sargisson.
274 pages. 2021. ISBN 978-1-78997-880-3

Volume 23 and 24	Darko Suvin: Parables of Freedom and Narrative Logics: Positions and Presuppositions in Science Fiction and Utopianism. Edited by Eric D. Smith (VOLS I & II) 706 pages. 2021. ISBN 978-1-80079-047-6 (set)
Volume 25	Valentina Romanzi: American Nightmares: Dystopia in Twenty-First-Century US Fiction. 304 pages. 2022. ISBN 978-1-80079-715-4
Volume 26	Lyman Tower Sargent: Rethinking Utopia and Utopianism: The Three Faces of Utopianism Revisited and Other Essays. 434 pages. 2022. ISBN 978-1-80079-489-4
Volume 27	Mónica Martín: The Rebirth of Utopia in 21st-Century Cinema: Cosmopolitan Hopes in the Films of Globalisation. 240 pages. 2023. ISBN 978-1-80079-442-9
Volume 28	Alexander Popov: Zone Theory: Science Fiction and Utopia in the Space of Possible Worlds. 368 pages. 2023. ISBN 978-1-80079-438-2
Volume 29	Pekka Kilpelainen: Postcategorical Utopia: James Baldwin and the Political Unconscious of Imagined Futures. 326 pages. 2023. ISBN 978-1-80079-233-3
Volume 30	A. L. Morton: The English Utopia. Edited with an Introduction by Antonis Balasopoulos. (Ralahine Classic) 302 pages. 2023. ISBN 978-1-78997-418
Volume 31	Pavla Veselá: The Polyphony of Utopia: Critical Negativities Across Cultures from Bellamy and Bogdanov to Yefremov, Piercy and Butler. 322 pages. 2024. ISBN 978-1-80374-055-3
Volume 32	Donald Morris: Economic Inequality: Utopian Explorations. 368 pages. 2024. ISBN 978-1-80374-176-5
Volume 33	Heather Alberro: Terrestrial Ecotopias: Multispecies Flourishing in and Beyond the Capitalocene. 304 pages. 2024. ISBN 978-1-80079-576-1

Utopian Studies

JENNIFER WAGNER-LAWLOR, EDITOR

Utopian Studies is a peer-reviewed publication of the Society for Utopian Studies that presents scholarly articles on a wide range of subjects related to utopias, utopianism, utopian literature, utopian theory, and intentional communities. Contributing authors come from a diverse range of fields, including American studies, architecture, the arts, classics, cultural studies, economics, engineering, environmental studies, gender studies, history, languages and literatures, philosophy, political science, psychology, sociology, and urban planning. Each issue also includes reviews of recent books.

ISSN 1045-991X | E-ISSN 2154-9648
Triannual | Available in print or online

Current pricing:
psupress.org/Journals/
jnls_utopian_studies.html

Submissions:
editorialmanager.com/uts

PENN STATE UNIVERSITY PRESS
www.psupress.org
journals@psu.edu

www.ingramcontent.com/pod-product-compliance
Ingram Content Group UK Ltd.
Pitfield, Milton Keynes, MK11 3LW, UK
UKHW021314180426
11947UKWH00015B/1223